THE VORTEX
THEORY

THE VORTEX THEORY

A Bridge between Ancient Yoga and Modern Physics

by

David A. Ash

Published by
Kima Global Publishers,
50, Clovelly Road,
Clovelly
7975
South Africa

© David Ash 2015

ISBN: 978-1-920535-73-5 this edition:

ISBN e-book edition 978-1-920535-72-8

Kima Global web site: www.kimabooks.com

Author's web site: www.davidash.info

Other books by David Ash:

The Vortex: Key to Future Science with Peter Hewitt (Gateway 1990)

The New Science of the Spirit (College of Psychic Science 1995)

Activation For Ascension, (Kima Global Publishers 1995)

The New Physics of Consciousness (Kima Global Publishers 2007)

The Role of Evil in Human Evolution (Kima Global Publishers 2007)

Continuous Living in a Living Universe (Kima Global Publishers 2015)

Acknowledgements

I thank my son Sam for encouraging me to write this book and Dr Manjir Samanta Laughton, for writing the foreword and including me in her science documentary. I have Anne McEwen Founder of the Essenes International, to thank for friendship and support in getting this version published and Suzie for her love and thoughtful encouragement as I wrote it. I have the 'Mystic Marquis' to thank for the 'figure of eight' clue that led me to understand nuclear energy.

My mentor the late Sir George Trevelyan and co-author the late Peter Hewitt lifted my work from obscurity and achieved its original publication; for which I have Aleck Bartholomew of Gateway Books to thank. Christiana of *Art of Life Films* and Tonya, thank you for the interviews you filmed. Priscilla and Robin Husband I am ever grateful to you both for your ongoing generosity of spirit, your care and your hospitality. I thank Steven and Renata for editing and Robin Beck, for publishing this and previous titles. Anna, I am grateful as ever to you for your ongoing friendship, kindness and charming drawings of elephants, ants, tadpoles and hedgehogs and Becky thank you for Mr. Proton. I have my mother to thank for her love, care, hospitality and her support and finally I wish to express gratitude to my father the late Michael Ash for introducing me as a small boy to the wonders of physics, for encouraging me by example to challenge the status quo, for teaching me to think independently and tread the lonely path of the pioneer but mostly for showing by life example that needs are met by faith that the Universe will provide, not by worry or concern.

Table of Contents

Foreword

No doubt historians existing in the late 21st Century and beyond will look back to these years spanning the second millennium as a major turning point in the history of science and in human knowledge. We are currently living through this revolution and, without the benefit of hindsight, the picture can seem muddy as there are still many detractors alive today. But the wheels are in motion and the revolution is upon us; our descendants will thank us for the efforts we have made in moving from the Newtonian, reductionist worldview of science, to one that incorporates consciousness at its core.

I strongly believe that one of the key protagonists who will stand out when this rear view mirror is used by future generations is David Ash. In the face of sometimes hostile ridicule from those who wish to perpetuate Newtonian dogma, he has had the courage to repeatedly point out where, in the current paradigm, the emperor has no clothes. He has the insight to ask the questions that no one else dares or even articulates that maybe our current scientific models are based on flimsy non-fundamental principles, supported more by historic precedent and academic funding than Nature.

So amongst the crumbling of the old, there are some like Ash who are bringing in the new. His vortex theory has the insight to solve many of the issues in current science and he does this with rigorous logic and humour, making his ideas accessible to the scientist and non- scientist alike. Indeed the theory has had the track record of predicting future scientific observations – a strong sign of its validity.

Ash disregards the prejudice of our current Western culture that sees ancient Eastern wisdom as somehow irrelevant and primitive and in doing so revives the true nature of wisdom: beyond the laboratory of modern science the universe is available for direct experience. In fact he knows that although our experiments using the scientific method have indeed led us to great knowledge, the wisdom of ancient cultures can be deeper, as it is obtained from within.

You have an incredible adventure in front of you during which you can leave behind preconceived divisions between aspects of knowledge. Ash will take you far and wide on this epic journey – from deep into the atom, to meditating in India, to joining Alice in Wonderland.

Crucially, he is one of the guides of this new revolution, shaking us up to look deeper and realise that the universe is as it is: it is we who have compartmentalised knowledge and created dogmatic rules. So let him take you back to the truth – that all knowledge exists in unity from East, West, macro to micro – there is no division. Let the revolution begin!

Dr Manjir Samanta-Laughton. MBBS, Dip Bio-Energy
Derbyshire, United Kingdom

October 2011

Introduction

I am hopeful we will find a consistent model that describes everything in the Universe. If we do that, it will be a real triumph for the human race.

Stephen Hawking

A book I read in my youth influenced the course of my life. The *Advanced Course in Yogic Philosophy*[1] by William W. Atkinson writing as Yogi Ramacharaka led me to the vortex theory. I came across the book in 1964 when I was still at school. As I read it I realised it contained a clue to understanding the way energy forms matter which could revolutionise physics and make it understandable to everyone.

My father, Michael Ash (1916-1991) introduced me to physics when I was seven so I knew a bit about Einstein. When I read the book on Yogic philosophy published in 1904, I was galvanised. It contended matter to be a form of energy a year before Einstein published $E=mc^2$. Einstein anticipated by Yogis! The thought I had was, "Yogis said energy forms matter through vortex motion. If they could anticipate Einstein maybe they were right about the vortex!"

That thought launched me on a voyage of discovery that has gone on for half a century. As I worked on the vortex theory no door in physics was barred to me; no mystery seemed impossible to solve. An account for the forces and particles of nature emerged until a complete panorama of the Universe lay before me. Most incredible was the conclusion in line with the consensus of many quantum theorists and the philosophy of Yogic mystics that there is a universal mind, a universal consciousness underlying everything.

My father was an experimenter and emphasised the need for a prediction that would enable my vortex theory to be tested by experiment. Sadly he died before the accelerating expansion of the Universe, predicted implicitly in my vortex theory, was confirmed by a discovery in astronomy.

In 1990 Saul Perlmutter, professor of Physics at Berkley University, pulled together teams in the USA and the UK to look for supernova explosions in distant galaxies. They published their results in December 1997.[1] Prior to then it was assumed from the big bang theory that the Universe was expanding at a uniform rate. Perlmutter's observations of exploding supernovas suggested this assumption was wrong and in fact the expansion of the Universe was accelerating.

To quote Martin Rees: "An acceleration in the cosmic expansion implies something remarkable and unexpected about space itself: there must be an extra force that causes a 'cosmic repulsion' even in a vacuum. This force would be indiscernible in the Solar System; nor would it have any effect within our galaxy; but it would overwhelm gravity in the still more rarefied environment of intergalactic space. Despite the gravitational pull of the dark matter (which acting alone would cause a gradual deceleration), the expansion could then actually be speeding up."[2]

I didn't know about Perlmutter or his study when I was working on the cosmological implications of my vortex theory that predicted a polar opposite to gravity. I could see that on a cosmic scale vortex energy could cause galaxies further from us to accelerate away from us faster than those that are closer. My account

1   **Perlmutter** Saul., et al. Discovery of a Supernova
     Explosion at Half the Age of the Universe and its
     Cosmological Implications Lawrence Berkeley National
     Laboratory, Dec 16, 1997
2   **Rees** Martin *Just Six Numbers,* Weidenfield & Nicolson
     1999

was published in 1995[3] two years before Perlmutter published his discovery of the accelerating expansion of the Universe, caused by *dark energy*, in 1997[4].

According to the rules of science if a theory can predict the outcome of a future observation or experiment it should be taken seriously, no matter how unknown the author or obscure the publication. There was little hope of people taking me or my work seriously unless I could establish the validity of the vortex theory according to the scientific method. Despite the fact that I had neither PhD nor peer support, the vortex gleaned from Yoga enabled me to explain things that baffled professional physicists. For example, I was able to account for potential and nuclear energy; how energy is locked in matter. In 1905 Einstein, without a degree in physics, published his equation $E=mc^2$, predicting that vast amounts of energy could be released from small amounts of mass. That was proved in the atomic bombs of 1945 but as the physicist Richard Feynman pointed out: "Nuclear energy... we have the formulas for that, but we do not have the fundamental laws...we do not know what it is."

The mystery of how subatomic particles release the explosive force of the nuclear bomb can be explained without difficulty by the vortex of energy. This book, an update of my 1995 publication, retraces how the vortex idea from Yogic philosophy can explain not only nuclear energy but many other enigmas in physics. If a theory can explain things that are unexplained, as well as predict the outcome of future experiments, by the rules of science it is a good theory.

3   **Ash D.** *The New Science of the Spirit.*College of Psychic Studies 1995

4   **Perlmutter** Saul., et al. Discovery of a Supernova Explosion at Half the Age of the Universe and its Cosmological Implications Lawrence Berkeley National Laboratory, Dec 16, 1997

Introduction

A sound scientific theory coming from the mystic tradition of yoga may be anathema to secular materialists, especially when developed by an amateur, but prejudice does not figure in science. If a theory complies by the rules of the scientific method, regardless of where it comes from or how inconvenient it may be it cannot be dismissed unless it is disproved. It can be ignored but that leads to ignorance!

Rules are rules. Motorists may object to the rules about parking and speeding but they have to follow them no matter how annoying and inconvenient they may be. A new scientific theory can emerge through anyone and at anytime, that not only challenges accepted beliefs but causes theories held true for centuries to be discarded. This can happen if the new theory explains just one fact that the old theories could not; as happened at the turn of the twentieth century when Max Planck resolved a problem with the way heat and light are radiated that could not be explained by classical wave theory. His suggestion that and heat were radiated in packets of waves enabled Einstein to explain another problem called the 'photoelectric effect', which led to his 1921 Nobel Prize and the foundation of Quantum theory.

The rules of science require us to accept a more simple theory that explains things with minimal assumptions over a complex theory that heaps speculation upon speculation. Economy of ideas is a hall mark of a good theory. Mass can be accounted for by *Higg's Boson* but the vortex theory for mass as spin is far simpler and more straightforward than Peter Higgs complex theory!

A scientific theory is allowed an initial assumption, called an axiom. A good theory will explain what is already known from its axiom then offer a prediction that can be tested by experiment. That is the scientific method. The vortex as an axiom enabled me, as a self taught physicist, to develop of a self consistent theory in physics which then enabled me to predict one of the most important scientific discoveries in the late twentieth century.

In science nothing is absolutely certain nor is science devoted to proving theories. In the scientific method new theories challenge the established ones and replace them if they are overthrown.

I stumbled on an amazing insight from the ancient mystic tradition of yoga and couldn't put it down. The scientific theory that has come from the thread I traced through physics speaks for the value of the insights of Yogic mysticism not me.

Stephen Hawking concluded *The Brief History of Time*[5] by predicting the discovery of a complete theory that would be understandable in broad principle by everyone and as a triumph of human reason it would enable us to know the mind of God. It would be a triumph for mysticism if a theory originating in Yoga fulfilled Hawking's prediction. The vortex theory is simple enough for everyone to understand and it does depict the Universe as a mind emerging from a single underlying Consciousness at the ground of all being. This could demonstrate that mystics can contribute to science, which might engender more respect for them than is reflected an attitude of the so called scientific enlightenment, caught in a comment by an Oxford academic: *"Only scientific criteria for truth are valuable and mystics are pathological cases."*

There are many ideas in the mysticism of Yoga that could be of value to humanity if we are open to them. They are profound and some could point to future directions of science; just as the idea of the vortex has done.

Yoga can interface with science better than religion because it does not require people to have blind faith in an exclusive system of belief. Science is never certain. In science there are no ab-

5   **Hawking** Stephen, *A Brief History of Time* Bantam Press 1988

solute truths, there are only hypotheses and theories. If sciencists were more tolerant of mysticism then the methods of scientific investigation could be applied to mystical insight; but only if scientists approach mysticism with an unbiased mind because if mind underlies everything then the beliefs of experimenters could influence the outcome of scientific experiments.

Yoga was and always will be more to do with the scientific principles of knowledge and experience than faith. In physics, the king of sciences, yoga comes into its full majesty. The principles of yoga in science could rescue us from dogmatic beliefs that allow for militant theocracy in religion and atheistic materialism in science.

Millions of people today embrace yoga and benefit from the exercises and meditation yoga has brought to the West. Now we have the opportunity to apply the wisdom of yoga to reconcile science and spirituality and lift our understanding of the Universe beyond limited assumptions and entrenched beliefs. Appreciating how energy spins into matter, to form us and the world in which we live, we can lift our sights to unforeseen levels and embrace a future full of possibility beyond our greatest expectations.

Chapter 1

The Vortex of Energy

I believe matter itself is just spin.

Eric Laithwaite

Lord Kelvin (1824-1907) was a towering figure in classical physics of the 19th Century. He pioneered thermodynamics but also the theory of the *vortex atom*. In Kelvin's day the atom was thought to be the smallest particle of matter but Kelvin despised the common assumption that atoms were solid particles like billiard balls. To him this model was unsatisfactory as it offered no explanation for the properties of matter. He considered the popular, materialistic view of matter as superficial and naïve and dismissed it, describing the billiard ball atom as: *"…the monstrous assumption of infinitely strong and infinitely rigid pieces of matter… Lucretius' atom does not explain any of the properties of matter…"* [1]

Kelvin picked up the vortex idea from the German physicist and physician, **Herman Helmholtz** (1821-94) who was convinced that atoms were vortices in the ether. In the Victorian era people believed the Universe was pervaded by the luminiferous ether which transmitted waves of light much as the ocean transmits water-waves. Kelvin taught that light and matter consisted of waves and vortices in the ether.

Thanks to the endorsement of Lord Kelvin, the theory of the vortex atom dominated physics in Victorian England in the latter

---

1   **Thomson**, W., *Mathematical and Physical Papers* 6 vols. 1841-1882

half of the 19th Century and continued to be taught at Cambridge until 1910.

James Clerk Maxwell (1839-71), who developed the math of electromagnetic theory, was a strong proponent of the vortex atom. In Encyclopedia Britannica of 1875 he wrote: *"...the vortex ring of Helmholtz, imagined as the true form of the atom by Thomson (Lord Kelvin), satisfies more of the conditions than any atom hitherto imagined ..."*.[2]

J. J. Thomson (1856-1940) who discovered the electron was professor of physics at Cambridge when he said *"... the vortex theory for matter is of a much more fundamental character than the ordinary solid particle theory."*[3]

With a vortex theory, physicists were able to reduce the properties of matter to a single, underlying, dynamic principle. But the Victorian vortex failed on two counts. Firstly it was applied to the atom which is not a fundamental particle. That was not the fault of the Victorian scientists. They had no idea that, in the century to come, the atom would be split. Secondly it was assumed the vortex was in the ether so when the ether theory was disbanded all the ether theories, including the theory of the vortex atom, were abandoned.

I first read about the vortex atom in *Advanced Yogic Philosophy* [4] by **William Atkinson** (1862-1932) writing as Yogi Ramacharaka. What impressed me in Atkinson's book was not his mention of the Helmholtz vortex ring but his reference to Yogis in ancient India appreciating that energy *prana* exists in matter *akasa* in the form of vortices; *vritta*. It was the Yogic prescient that fundamen-

---

2   **Maxwell** James Clerk, *Encyclopaedia Britannica*, 1875
3   **Thomson** J.J., *Treatise on the Motion of Vortex Rings* University of Cambridge 1884
4   **Ramacharaka** Yogi, An Advanced Course in Yogi Philosophy 1904, p 320

tal particles in matter were forms of energy that impacted me sufficiently to consider the vortex theory. Throughout the book Atkinson was emphatic that matter is a form of energy. The copy I was reading from a library of antique books was put to print in 1904, a year before **Einstein** (1879-1955) published his famous equation equating mass and energy. Atkinson spoke of the atom because in his day it defined the smallest particle of matter. In 1964 I realised it must have been a subatomic particle, not the modern atom, that the Yogis had described as a *vortex of energy.*

Yogis in the pre-scientific era

In subsequent reading I discovered references to Yogis in the pre-scientific era probing matter with a supernormal *siddhi* power. This practice by Yogis was recorded about 400 B.C.E. in the 'Sutras of Patanjali'[5] where the results of meditation were described in detail. In Aphorism 3.46 it states that through meditation the Yogi can gain the power from the practice of the *anima siddhi,* of shrinking their consciousness commensurate with the atom. Yogic knowledge of the vortex of energy in the atom came from direct perception. Mystics in ancient India observed subatomic matter through meditation. Yogis, through meditation actually saw *quantum spin.* The siddhi system in ancient Yoga, for direct observation of quantum reality, undermined my confidence in the uncertainty principle of Werner Heisenberg that underpins quantum mechanics.

I began to work seriously on the vortex theory in 1968 when I entered Queens University of Belfast. I was prompted by Professor, Gareth Owen to develop an original theory. He expressed concern that students were treating universities as extensions of school when they were intended to be places to explore ideas,

---

5   **Prabhavananda** Sri Swami, *The Yoga Aphorisms of Patanjali* Sri Ramakrishna Math, Madras & Vedanta Society California 1953

question accepted beliefs and develop new world views. Why he chose me to speak to on the importance of developing a theory I will never know, but he catapulted me on a mission to test my idea of subatomic particles as vortices of energy against known physics.

The first challenge I encountered at Belfast was to find a vortex without poles. That word *vortex* conjures up a picture of the conical spin of a tornado or hurricane or water swirling down the plug hole in a bath. However, any three dimensional spiral is a vortex and there are many forms of these. Kelvin used the toroidal vortex, depicted by a smoke ring, for his model. It was convenient because smoke rings lent themselves to his lecture demonstrations.

Like a Ball of Wool

A lecturer in physics asked me how subatomic particles could be vortices when they were corpuscles without poles. My breakthrough came when I watched someone wind wool into a ball. I saw that when the wool was wound about a single point its freedom to wind in every direction caused to form a ball. As the ball of wool grew in font of me I saw it as a dynamic vortex.

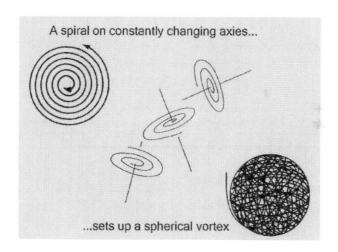

A spiral on constantly changing axies...

...sets up a spherical vortex

A conical vortex spins about an axis so it has poles. A ball of wool is a spherical vortex. A spherical vortex could be a corpuscle without 'discernable' poles if the axis of spin continually changed orientation. If the energy in a spherical vortex was spinning 'at the speed of light' the poles of spin could be changing too fast to measure at any moment. I proposed spin on changing planes, setting up a corpuscular subatomic vortex of energy, as the fundamental subatomic particle of matter. This I described as the *quantum vortex;* a literal form of quantum spin.

The spherical vortex of energy provided me with a working vortex model for subatomic particles of matter such as electrons, protons and neutrons as corpuscles without poles. Kelvin's vortex ring would not work for subatomic particles. He would have been floored by that physicist from his city of birth. For a model to work in physics it has to satisfy observations. As the years went by I found my vortex model worked well with experimental physics and it was also convenient. Whereas Kelvin had to cart cumbersome smoke boxes around to his lectures, along with an assistant to operate them, all I had to do was remember to take a ball of wool along with me to my lectures and if I forgot, there was always someone or a shop nearby with wool that I could wind frantically in front of my audience.

In Belfast I proposed the vortex as the fundamental 3D form of energy that conferred three dimensions onto matter. We live in a three dimensional world and it is easy to take *that* for granted. The spherical vortex of energy provided a reason *why* matter extends in 3D – three dimensions.

Inertia

Another property of matter is . **Richard Feynman** (1918-88) one of the greatest physicists of the 20th century, said that: *"The laws of have no known origin."* [6]

The vortex of energy offered me a simple account for . A gyroscope is a horizontal spinning flywheel. A massive gyroscope acts against the list and pitch of a ship in an angry sea to help keep it level. A gyroscope demonstrates how spin sets up . A spinning pebble skipping across the surface of a pond also illustrates the set up by spin. The gyroscopic spin of the pebble keeps it in the plane in which it is thrown so that it skims rather than sinks.

I reasoned that in a spherical vortex the simultaneous spin of energy in all directions would set up *static* , a resistance to movement of the vortex in any direction. This is how I imagined the subatomic vortex setting up the of mass. I accounted for in terms of spin.

I defined mass as the quantity of vortex energy in a particle of matter. When I started the vortex theory in the 1960's there was no talk of Higgs bosons. The vortex theory provided me with a satisfactory account for mass based on spin, which was a lot easier to understand than Peter Higgs' more recent account in quantum mechanics. My theory has the advantage of simplicity and *economy of ideas.*[7]

Energy is motion. Motion creates . This is illustrated by balance on a bicycle or skis. The forward motion on skis or bicycles sets up resistance to sideways movement out of the plane of motion. So it is in the atom, the spin of electrons and their motion in or-

---

6   **Feynman** R. *The Character of Physical Law*, Penguin
      Books 1997
7   **Ariew R.,***Ockham's Razor* University of Illinois Press 1976

bits contributes to the inertia of matter. Albert Einstein's original paper equating energy with mass did so in terms of inertia.[8]

Einstein had the equation $E=mc^2$ in that epoch making paper but he lacked a simple and straightforward model to help people visualise matter as energy. The vortex shows how energy can form mass. It accounts for inertia and the extension of matter in three dimensions. The vortex of energy explains away the proerties of material as a delusion set up by spin.

Belief in material Substance

In Indian philosophy belief in material substance is treated as an aspect of perceived delusion called *maya* the 'illusion of forms'. Vortex motion underlying matter sets up the maya of 3D form characteristic of matter; the inertia of mass, the 3D extension of matter and the corpuscular form of fundamental particles. Through the vortex of energy I came to appreciate the material delusion; the way the vortex can hoodwink humanity! The fundamental properties attributed to material substance are properties of the vortex of energy. People who endeavour to explain away the non-material are deluded by vortex energy into belief in material substance.

Lord Kelvin described the 'billiard ball atom' as a monstrous assumption because he realised the concept of material particles with irreducible properties was a delusion set up by vortex motion. When I read a book on worldviews by William Berkson I realised Einstein was not deluded by material substance. William Berkson said Albert Einstein was difficult to understand, not because of his theories and his mathematics but because of

---

8   **Einstein**, Albert, *Does the Inertia of a Body Depend Upon Its Energy Content?* Annalen der Physik 1905

his worldview: *Einstein believed matter and the field are real but there is no substance to them.* [9]

In 1905 Einstein predicted that mass is relative to the speed of light. He equated mass to the speed of light in $E=mc^2$. The speed of light is a measure of movement and it struck me that particles of energy were particles of motion at the speed of light. Through the vortex I was able to explain *how* particles of movement at the speed of light could form particles of matter. I saw *how* energy can be equated to mass. When I was teaching my vortex theory I developed a line of logic to explain how mass came from motion not material.

Rain in the Alps

Imagine rain in the Alps. As water falls to form streams tumbling down the mountainside, these join the torrent of rivers that pass through hydroelectric plants. There the fall of the water is converted into the spin of turbines and then the flow of electricity. The electricity is then fed into CERN where it is used to accelerate protons. As the protons collide in intersecting rings of the particle accelerator the motion is arrested. The arrested kinetic energy is then transformed into the mass of newly created particles.

Would anyone suggest that in the fall of rain, and the tumble of streams, the torrent of rivers and the spin of turbines, the flow of electricity and the acceleration of protons some mysterious material substance was transmitted to be transformed into the new particles? Motion alone was transferred from one step to another. It is obvious that the newly formed particles of matter had to be forms of motion.

9   **Berkson**, William, Fields of Force: World Views from
     Faraday to Einstein, Rutledge & Kegan Paul 1974

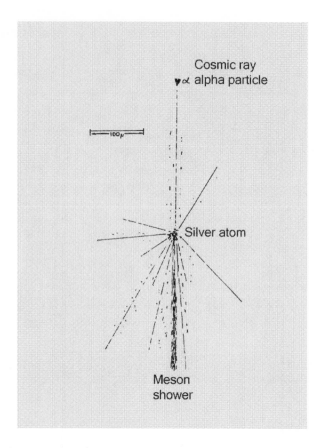

A year after I discovered the vortex of energy I saw how the insight of Yogis could account for the transformation of motion into mass. It was September 1966 and I was nearly eighteen, just beginning my A level GCE course in physics at the Hastings College of Further Education. I was browsing a new text book[10]. The last plate portrayed a high energy event witnessed by **Professor Cecil Powell** (1903-1969) of Bristol University in a photographic

10   **McKenzie** A. E., *A Second MKS Course in Electricity*, plate 19, Cambridge University Press 1968

plate he sent to high altitude in a weather balloon, in his cosmic ray research.

Powell's photograph witnessed a cosmic ray particle from the sun smash into the nucleus of a silver atom in the photographic emulsion. A shower of newly created particles called *mesons* appeared from the collision. The caption under the photograph said...*this is a vivid demonstration of the transformation of kinetic energy into mass.* [10]

The cosmic ray experiment was proof enough for me that mass was a form of movement. If the mass of the meson particles could be created out of the arrested motion of the cosmic ray particle then all mass in the universe could be a form of movement and that form I believed was the vortex.

I reasoned, when the cosmic ray particle was stopped its kinetic energy continued through the atomic nucleus. This *motion* was forced through the proton and neutron vortices in the nucleus. Like batter forced through a doughnut mould the kinetic energy took on the vortex shape, much as the batter would take on the doughnut shape. The new vortices of energy then appeared on the other side of the nucleus as the shower of mesons captured on the photographic plate.

Powell's cosmic ray picture provided me with experimental evidence that motion is the basis of mass. The picture also revealed that energy forming matter is pure movement. Years later in a book by Nigel Calder on the new physics I read a quote by Richard Feynman: *"It is important to understand in physics today we have no idea what energy is."*[11]

11   **Calder** Nigel, Key to the Universe: A Report on the New Physics BBC Publications 1977

What is Energy?

Thanks to the Yogic vortex it was clear to me what energy is. Energy is activity; it is pure movement. This was reinforced years later when, as a student at London University, I read Fritjof Capra's book *The Tao of Physics*. Because my vortex physics was derived from the Eastern tradition I was intrigued by it. Capra's second book *The Turning Point* confirmed my understanding of energy as motion. I realised it wasn't just the Yogi's who appreciated the basis of our world is movement, not material. The dynamic underlying everything was the understanding of the Vedas and the Chinese sages and Buddha himself realised there is nothing in our world but changes, processes and ceaseless activity: *"There is motion but there are, ultimately, no moving objects; there is activity but there are no actors; there are no dancers, there is only the dance... Like the Vedic seers, the Chinese sages saw the world in terms of flow and change, and thus gave the idea of a cosmic order an essentially dynamic connotation... The general picture emerging from Hinduism is one of an organic, growing and rhythmical moving cosmos; of a universe in which everything is fluid and ever changing, all static forms being maya, that is, existing only as illusory concepts. This last idea – the impermanence of all forms – is the starting point of Buddhism. The Buddha taught that all compounded things are impermanent', and that all suffering in the world arises from our trying to cling to fixed forms – objects, people or ideas – instead of accepting the world as it moves and changes. The dynamic worldview lies thus at the very root of Buddhism."*[12]

Philosophers in the Far East were not duped by materialism. They did not succumb to the material delusion. They did not assume the existence of the ether or solid atoms. They recognised

---

12   **Capra** Fritjof., *The Tao of Physics*, Wildwood House, 1975 & *The Turning Point*, Fontana, 1983

the dynamic state underlying everything; the ceaseless movement that we in the West call energy.

Scientific Materialism

A problem we are faced with in the West is the material assumption originating in ancient Greece. It arose from the speculations of philosophers like Democritus (470-360 BCE) and Aristotle (384-322 BCE). The idea Democritus promulgated was that matter consists of atoms moving in a void of space and that nothing else is real. He was vehemently opposed to belief in soul and spirit and his ideas led to the philosophy of scientific materialism. His material view of the universe became the cornerstone of science. Richard Feynman put this very succinctly: *If in some cataclysm, all of scientific knowledge were to be destroyed, and only one sentence passed on to the next generation of creatures, what statement would contain the most information in the fewest words? I believe it is the atomic hypothesis... that all things are made of atoms - little particles that move around in perpetual motion."*[13]

The atomic doctrine of Democritus, summed up as *little particles that move around in perpetual motion,* didn't allow for the motion of a cosmic ray particle to transform into particles of matter or for them to revert to radiant energy! In the vortex hypothesis particles of kinetic energy can *transform* into vortices which can unravel back into wave motion again.

Democritus based his philosophy on speculation. His idea of fundamental particles was correct but he didn't have the *siddhi* power of the Yogis to perceive those particles as vortices of energy. The atomic hypothesis came from one strand of antiquity. The vortex came from another. Bringing them together could lead us to a better and more complete understanding of the Uni-

---

13　**Calder** Nigel, Key to the Universe: A Report on the New Physics BBC Publications 1977

verse. In our multicultural society we have access to ideas from every quarter of the globe. The question is, are we open to receive them? Will we ever come to revere the philosophers of ancient India as we revere the philosophers of ancient Greece? In 1966, I captured this sentiment in the lyrics of the first song I composed:

To the Diamond of Truth there are countless facets,

Few men can see more than one,

The seeker of truth sees just one facet,

And of the diamond he thinks it's the sum,

The light of each facet is different,

Each has its own tale to tell,

But in essence the light of the diamond

From each facet is always equal.

Facets of Truth

The philosophies of East, as well as West, represent different facets of the diamond of truth, they complement each other. They both played their part in my discovery of the vortex model of subatomic matter. I also saw Einstein and Newton as complementary. From the physics of Sir Isaac Newton (1642-1727) I came to understand the laws governing the transformations of energy between vortex and wave. These eventually appeared in my work as the *quantum laws of motion*. My account for Einstein's equivalence of mass and energy was based on Newton's laws of motion. If everything in the Universe is energy and energy is pure motion, it made sense to me that laws of motion should be fundamental in physics. That was the genius of Newton. For me Newton was not usurped by Einstein; they were different brilliant facets of the same diamond of truth that all is motion.

Chapter 2

The Quantum Laws of Motion

...this firm belief in a superior mind that reveals itself in the world of experience, represents my conception of motion

Albert Einstein

I wasn't taught physics at my secondary modern school so when I applied myself to the subject at college I approached it with an unconditioned mind. I was old enough to question what I was taught and I was reared on original thinking.

From my boyhood study of Einstein I knew energy was the stuff of everything. In my teens cosmic ray research provided experimental evidence that mass was based on movement and not material substance. By revealing the delusion of material, the vortex enabled me to understand how movement could form mass and how in cosmic ray research kinetic energy passing through the nucleus of an atom could be transformed into particles of matter.

Powell's particles were exceedingly short lived. They decayed in a thousand trillionth of a second so as soon as the newly birthed vortices escaped the nucleus they vanished.

Years later, when I gave a talk in a Sydney prison, I likened particles of kinetic energy to the prisoners and the nucleus to a prison. The prisoners were only such whilst they were in prison. As soon as they escaped or were released they reverted to free people. Likewise the particles of kinetic energy were only spinning while they passed through the vortex particles in the nucleus. As soon as they broke free of the nucleus they reverted to their original free radiant form. In my analogy humans represented energy and the state of being imprisoned or being free represented two possible states of energy. People are still people whether they are imprisoned or free. So it is with energy. Energy is neither created nor destroyed. It is always the same essential reality. It is just the states or forms of the energy that can change.

The Vortex Model

The vortex model enabled me to understand how energy formed particles of matter but I didn't have a model for the form of energy in light. I needed a picture for the kinetic energy radiating away as the mesons escaping the nucleus of the atom decayed. I needed to discover the laws governing transformations of energy between matter and light, but couldn't do that when I was busy swatting for my A level's in physics, chemistry and biology and at home my father had set me the task of securing him a US patent. That took up a lot of my time, not to mention keeping his accounts. Coping with the constant stream of diversionary maelstroms that popped out of his head made it difficult for me to think about the vortices in mine.

The wrath of Whitehall that descended on us, when he launched an alcohol antidote to beat the breath test was an overwhelming distraction. Along with helping my brother solder his diagnostic instruments on the kitchen table I had minerals to mix on that same board for then as now I was between nutrition and physics. With my brother Simon I was also learning the guitar. I couldn't focus on my vortex theory until I escaped to university in Belfast. At college all I managed was to develop a model for the radiant form of energy.

I never liked the classroom definition of 'kinetic energy'. The teaching in my O-level GCE (GCSE) year was grounded on the material delusion that movement couldn't exist without something that moved. It didn't allow for the movement of a cosmic ray particle to continue after the moving particle had stopped and it certainly didn't allow for that unsupported movement to transform into mass!

I knew that energy existed in particles. That was the core of quantum theory and I realised particles of energy were not 'material things' but I lacked a conceptual model of the *immaterial quantum.*

In my first year at college I befriended an out of work actor called Robin Piper. He was a mature student with a passion for language. He always carried an etymological dictionary and during breaks between lectures or over lunch he encouraged me to look up the origins and meanings of words. That is how I first came to appreciate energy as *the movement within.* Robin's other passion was the abstract. As an actor he worked in a world of fantasy where the imagined became reality. Robin helped me to grasp the concept that the abstract could be real. He quoted Shakespeare that life is a play. He added we are but actors performing our parts and nothing need be real in the material sense if everything were real in the way that the plays of Shakespeare were real. I lost contact with Robin when I started my A levels but his vision of abstract reality prepared me for the epiphany that greeted me when I first glanced at the cosmic ray photograph.

The Dream State

Robin introduced me to the Aboriginal concept of the *dream state.* He laughed that they were far in advance of civilised Europe. They appreciated life was a play long before Shakespeare. He started me thinking that particles of energy might be particles of abstraction, bits of imagination, dream fragments, acts without actors. Robin's influence led me to conclude that *particles of energy are more like thoughts than things.* Robin didn't believe in God whereas I was brought up a Catholic so there was a lot we didn't agree on but we met on the possibility that the Universe could be a mind. I said it was the mind of God. He replied I could not assume the existence of God separate from the Universe because as a mind the Universe could be God! He was OK with the idea of God if God was the Universe; if everything was God. He pulled out his dictionary and got me to look up *pantheism.*

I needed a model for energy as motion. I imagined it as a line of movement. I envisioned the lines spinning to form vortices in matter or undulating in waves to form light. Later I would use wool to depict these lines. I didn't know about the string theory that was being developed on the West Coast of America. It was years before I realised I had developed a string theory in England quite independently. They spoke of strings of vibration. I saw strings of spin as well as vibration.

I determined the Universe was formed out of little lines of movement at the velocity of light spinning or vibrating. The lines were not substantial things; they were events at the speed of light, which were fundamental to everything. All that was consistent about these bits of movement was their shape and their speed. That was how I understood Einstein when he said the speed of light is the sole universal constant. Einstein said everything is relative to the speed of light. I argued everything was made of minute bits of activity at the speed of light; little whirlpools of spin or bundles of vibration happening at the speed of light.

Cosmic Rays

Cosmic ray research proved to me that movement not material was the fundamental reality. The movement of the cosmic ray particle that continued through the nucleus when the particle had stopped was a line or string of vibration. It was a line of wave form kinetic energy that could transform into spin if it was forced through a vortex in the atomic nucleus. It then reverted to vibration when it escaped. It was the same 'string' or line of the movement of light throughout.

At college I drew a diagram to illustrate the line of the movement in light in wave form.

My diagram represented a particle of energy as a line of motion at the speed of light. The amount of energy was represented by the length of the line. During my A level studies we were taught that energy occurred in *quantum* bundles. That description helped me because when I folded a line into a bundle, the greater the number of folds, the greater was the length of line. The folds in the lines represented waves. My model showed why the energy in a quantum is proportional to the frequency or number of waves it contains.

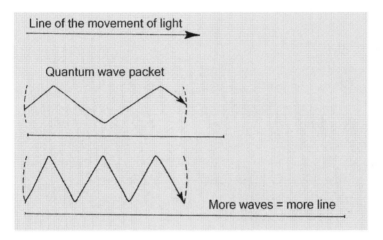

Years later I considered the bundles of vibrating lines of energy driving the cosmic ray particle on its long journey from the sun before it collided with the nucleus of a silver atom in Professor Powell's photographic plate. I conjectured that when the particle was stopped in the crash the lines of movement just went on, passing through the vortex particles in the nucleus, where they were transformed from undulation into spin; the energy was changed from light into matter.

Appearing on the other side of the nucleus the lines didn't vanish; they merely reverted into the kinetic form that wasn't detected in the experiment. The undetected waves were the original form of the kinetic energy. If they were reverting to their original form it struck me that the particles of energy must have had memory of their original shape. After the form was forced to change it sprang back to its original shape as soon as the force was removed, and it happened very quickly, in less than a trillionth of a second! I thought about this over many years until two self evident principles crystalised in my mind which then led to my quantum laws of motion:

The motion of energy is perpetual.

The first law of thermodynamics is energy is neither created nor destroyed. If energy is motion then that motion must be perpetual. This idea does not fit with common experience. Usually work has to be done to keep things moving. As soon as we slack they slow and then stop. The idea of perpetual motion is anathema to scientists and normal human thinking but it is fundamental to the permanence of the Universe of energy. If nothing exists but the bits of movement we call energy, should the movement stop then everything would cease to be. The Universe would vanish! When the cosmic ray particle stopped, its motion could not stop. It couldn't just cease to be. It had to go on, through the nuclear vortices and out the other side. Einstein founded quantum theory and received the Nobel Prize by showing that quantum particles of energy are not temporary but are permanent.

The motion of energy has shape.

Scientists have discovered from their experiments and observations of the natural world that waves are a fundamental form of energy. Nobody has been able to explain 'why' energy occurs in waves. However, scientists throughout the world agree beyond shadow of doubt that energy 'does' flow in waves. This conclusion is fundamental to quantum theory. It is the basis of everything we know about light and sound. Astrophysicists assure us that the waves of energy released in the Big Bang still exist somewhere so it is fairly safe to conclude that energy occurs in the shape of waves and these are a fundamental and perpetual shape of energy.

Helmholtz and Kelvin believed that our world is formed of vortices as well as waves. The axiom of my vortex theory is that the vortex is a fundamental and perpetual state of energy. Years later when I read that protons have been estimated to last for at least a billion, trillion, trillion years[1] I realised if these are vortices of energy then my conviction, that the vortex was perpetual, was reasonable.

Newton's Laws of Motion lived on in my mind from my first year of physics. They were enshrined in my O Level physics text book which I kept for years[2]. If particles of energy are particles of motion I imagined Newton's laws of motion must apply to them.

Newton's First Law

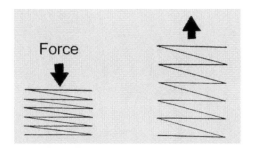

Newton's first law is that *something will maintain its state of motion or rest unless it is forced to change*. From that law I imagined each particle of energy as a tiny spring that would change shape only when forced to and if the force was removed it would revert immediately to its original shape.

These realisations didn't come immediately. In my teens embryonic ideas were gestating in my mind. Then decades later, after my understanding of nuclear energy was stimulated by the insight of a mystic Marquis, I formalised my understanding

1   **Calder** Nigel, *Key to the Universe:* BBC Publications 1977
2   **Abbot** A. *Ordinary Level Physics* Heinemann Education
1963

of energy mass transformations into two **Quantum Laws of Motion:**

The First Quantum Law of Motion:

A particle of energy will maintain its state of wave or vortex motion unless change is forced upon it.

The Second Quantum Law of Motion:

If the force of change is removed, the particle of energy will revert to its original state of motion.

Using these laws I was able to explain the cosmic ray experiment on the basis that kinetic energy forced into spin, as it passed through the vortex particles in an atom, reverted immediately to its original wave form as soon as it came out the other side of the nucleus.

To appreciate what was going on I applied Newton's second law of motion that *for every action there is an equal and opposite reaction.* Over the years it was obvious to me that the kinetic wave form of energy was opposite to the static vortex form. My father likened the interactions of wave and vortex to marriage where an irresistible force meets an immovable object!

I was taught that Newton's laws were the laws of inertia and I knew Einstein's equation governing energy-mass transformations, $E=mc^2$ was published in a paper on inertia. My father's quip brought it all together in my mind. There were two opposite forms of inertia set up by energy. One was static and the other kinetic and the tussle between these two opposing inertias set up *quantum dynamics*; the interacting and opposing actions and inertias of energy.

My thinking was simple. The vortex of energy set up the static inertia characteristic of matter. The wave train of energy set up the kinetic inertia characteristic of heat and light. I imagined whirlpools of energy staying put because of motion in spin. I envisioned flying bundles of energy constantly moving because of radiant wave motion.

The quantum laws helped me to explain *entropy*. I understood entropy as the tendency of particles of energy to maintain the inertia set by their shape or state of motion.

I thought of vortex particles of matter as passive inertia and wave kinetic particles of heat and light as active inertia. I saw in energy the power of opposites. To me it was elegant. One shape of energy set up a state of activity opposed to another form setting up a state of rest. It was simple, the entire Universe worked on the active versus the passive. That was to become the core of my new approach to quantum theory.

The Active Inertia of Light

The active inertia of light originates from the flow of energy in a train of waves. Waves of light will radiate endlessly. They never tire or run out of steam. This is because they are particles of energy in a perpetual form of kinetic or propagational motion. The active inertia of light is the *nature* of a wave-particle of energy to keep moving in waves unless it is stopped. The passive inertia of mass, set up by the vortex of energy, is quite the opposite. It is the *nature* of a particle of energy to stay put unless it is forced to move.

I understood wave-particle duality as the interplay of opposing states of inertia originating in opposite forms of energy. The dance between them was the cause of *work* through which everything useful and worthwhile happens. I realised the working of the Universe and life itself depends on the endless tussle between the active and passive forms of energy; of kinetic and static states of inertia. All this depended on the memory of the shape of the energy.

Someone asked me how particles of energy could have memory. I replied that a particle of energy did not *have* memory – in the materialistic sense of something existing to possess properties. It is more a particle *of* properties than a particle *with* properties.

Particles of energy could not *possess* memory; they could only *be* memories.

Universal Mind

The Yogis believe in Universal mind which they call *chitta*[3]. They believe that energy or *prana* is an expression of *chitta*. Yogis treat energy as an expression of thought. If the quantum laws of motion operate on the principle of memory and particles of energy are not things *with* memory but *are* memories there must be a Universal mind because memories are an expression of mind. If particles of energy are memories the Yogic idea of a Universal mind is reasonable. Just as we write or paint to record our thoughts, lest they be lost forever, so matter could be the canvas for a Universal mind; it could be a memory of creative thought.

The quantum laws of motion led me to conclude that mind is universal. The Yogis spoke of a single source of the universal mind. Some might say that is God. In my book the source of the Universe could not be God if God is defined as spirit, as spirit is just another dimension of energy. When I read *The God Delusion* by Richard Dawkins, I concluded humanity is more deluded by belief in material than by belief in God. If the quantum laws of motion reveal energy to be memory, as that infers mind those who choose to call it God would not be deluded in their beliefs.

Skeptic atheists claim that science refutes belief in God but I am convinced that this view does not represent true skepticism. A true skeptic will keep an open mind and think for him or herself. A true skeptic has the courage to continually question his or her own beliefs; not just everyone else's. A true skeptic doesn't just accept what is taught in science; they will challenge science as well religion and question belief in material as well as faith in God. Most people who call themselves skeptic today are following the tramlines of skepticism established during the scientific *enlightenment* when free thinkers began to challenge accepted religious beliefs. Then science was used to dispel religious fear and superstition and anti-religious skepticism was an important force in the emancipation of humanity after centuries of intolerance and suppression of free thinking by established religions. But today people say they are skeptics, not because they think for themselves but because they have been taught at a State school rather than a Sunday school!

Most skeptics these days are anti-spiritual because they still believe the mechanical world view of classical scientific materialism. Materialism is not skepticism! Albert Einstein set a standard for true skepticism when he dispelled the doctrine of scientific materialism through $E=mc^2$.

Homeopathy

Take homeopathy for example. Homeopathy does not fit with the material paradigm but it works with the idea that energy and matter may be memory. Most people, even if they use homeopathy don't understand it. Skeptics denounce homeopathy on principle or are funded by the Royal Pharmaceutical Society. I am biased toward homeopathy because my father practiced it and he taught me to dilute and potentise homeopathic remedies. My father was convinced that homeopathy would be explained one day by science and encouraged me to become a scientist in order to explain it.

When I elucidated the quantum laws of motion I discussed with my father the possibility that an account for homeopathy might be derived from the idea that matter is memory rather than material. He was delighted with the idea. He said homeopathic remedies were made by vigorous mixing of medicinal substances in a substrate like water or sugar. He suggested maybe an impression of the substance was made in the substrate like footprints in the snow. As the homeopathic potion was diluted, by mixed it in more substrate, perhaps the imprint that had no side effects was increased while the quantity of medicinal substance that could cause side effects was decreased.

Doctors, he said, gave patients substances that were poisons in large amounts but medicines in small amounts. When he practiced homeopathy, he observed a homeopathic dilution of a medicine was as effective as the medicine itself. He suggested that when homeopathic medicines were diluted they became more potent because the poisonous chemical decreased while the memory, the healing imprint of it, increased. He liked the idea that matter was memory rather than material because he could see with the quantum laws of motion homeopathy might become more acceptable to science, despite obvious resistance of the drug firms. He wouldn't join the NHS because he said the Hippocratic Oath would not allow him to use drugs with dangerous side effects when he knew that homeopathic medicines were safer and could be just as effective.

Homeopathy is not used by most doctors because they think it is strange. However, that is not a valid reason to dismiss it as unscientific. Strangeness is embraced by science because some of the new particles discovered in high energy laboratories are strange and strangeness has come to be accepted as a fundamental property of matter. Just as they helped me to understand the strange medical practice of homeopathy, my quantum laws of motion enabled me to appreciate why some subatomic particles are strange.

Chapter 3

Strange Particles

Even for the physicist the description in plain language will be a criterion of the degree of understanding that has been reached.

Werner Heisenberg

Cosmic ray research was the beginning of high-energy nuclear physics but though innovative and inexpensive it was haphazard as the energy of cosmic rays could not be controlled so physicists designed particle accelerators in order to control and increase the energy of their bombarding particles. By passing increasing amounts of energy through atomic nuclei, physicists were able to create ever more massive particles. This came to be known as *the particle zoo*. The most significant feature of the menagerie of particles generated in high energy research was short lifespan. With the quantum laws of motion I was able to explain how the new particles were formed and why they decayed so rapidly.

My account for the short-lived particles was simple. They were merely temporary vortices of energy, d into existence as kinetic energy was squeezed through the natural vortices in the bombarded nucleus. In keeping with the first quantum law of motion, the wave form of energy was d into the vortex form as it passed through the nuclear vortices. In line with the second, rapid decay of the new particles occurred immediately after leaving the nucleus as the newly formed vortices sprang back to their original wave form. That occurred because spin was not their natural state of motion.

In high-energy experiments some of the man made particles were observed to decay within 10^{-25} sec (a trillion trillionth of a second). That is the time it took for light to traverse an atomic nucleus. Others, however, lasted much longer, persisting for as long as 10^{-10} sec (ten billionth of a second). Physicists were puzzled why some new particles would survive more than a million times longer than others. The American physicist, Murray Gell-Mann, author of the Quark theory, called these durables in the particle zoo, *Strange Particles* and *Strangeness* came to be known as a fundamental property of matter.[1]

Short-lived particles, manufactured in the high energy laboratories, are artificial, highly unstable vortices of energy. The quantum laws of motion require that these unstable vortices decay back into the kinetic, wave form of energy immediately. Any longevity of an unstable vortex particle is a challenge to the laws. There has to be something delaying the process of energy unraveling from the state of end spin.

A common feature of strangeness is that the strange particles were decaying into stable particles. They were leaving behind a proton or an electron and sometimes even both.[1] If the high energy manmade particles were transitional vortices of energy it struck me that a strange particle could occur if a transitional vortex formed around a stable naturally occurring vortex picked up in the passage of the energy through the nucleus.

Hailstones

I imagined a natural vortex *scaffolding* the unnatural swirl of energy spinning round it. I concluded the vortex from the nucleus would stabilise the swirl of unstable energy by helping to hold it in vortex spin. This stabilising effect could slow down the unraveling of the strange particle after it explodes out of the atomic nucleus. Hailstones provided me with an analogy for this process.

1   **Calder** Nigel, Key to the Universe: A Report on the New
      Physics BBC Publications 1977

A hailstone forms in a cloud as a result of layers of ice forming around a seed of dust. The hailstone then falls from the cloud and melts, layer by layer eventually leaving the particle of dust behind. As energy is driven through an atomic nucleus in a high-energy experiment, it could form like a hailstone spinning

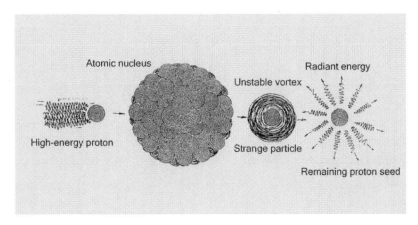

around a natural vortex such as a proton acting as a seed. It would then emerge as a new unstable vortex with a stable vortex at its core. On leaving the nucleus, the swirling energy could then decay like ice melting from a hailstone. Reverting from vortex to wave it would then radiate away leaving behind the stable particle, much as the hailstone melts to leave a seed of dust behind it.

My hunch that the proton vortex has a stabilising influence on the temporary vortex forming around it was clear from the way some strange particles shed their mass.

Particles appear in accelerator experiments at critical levels of mass-energy much as electrons occur in the atom at definite energy levels called *quantum states*. As physicists increase the energy in their accelerators they generate increasingly more massive particles at successive intervals. Some massive particles then decay in steps to reveal a new lighter particle at each lower levelof energy. The levels of mass-energy formed in a strange particle are referred to *doses of strangeness.*

Shedding Doses of Strangeness

The decay of a particle in steps, to reveal lighter particles, is described in physics as *shedding doses of strangeness.* If the stable proton, at the core of the unstable strange particle, were to exert a holding influence over it, then as each dose of strangeness is shed, the holding influence over the residual particle of transitional spin would obviously be stronger. This is because there would be less transitional vortex energy to stabilise. The stronger scaffolding influence on the remnant temporary vortex energy would also be due to the fact that it is closer in to the natural stabilising vortex core. Consequently as each dose of strangeness is shed the new particle should possess a longer lifespan than the one that existed before it.

In February 1964, at the Brookhaven National Laboratory in America, physicists witnessed the cascade decay of a heavy particle they called the *Omega minus.* [1] Looking closely at the bubble-chamber photograph of the cascade decay of the Omega

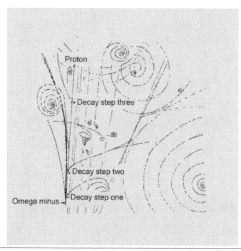

Cascade Decay of the Omega Minus particle drawn from a picture taken by Nicholas Samios & colleagues at the

minus I saw after each successive decay step the length of the track was longer. This showed that each successive particle had lasted longer than its predecessors. (Note the vortices in the bubble chamber photograph. Vortices are very common in pictures of high energy experiments.)

I traced the drawing from the original photograph. If you look carefully at it you will notice a successive increase in the track after each step in the process of cascade decay. That is a clear indication of increased longevity. This not only supports the quantum laws of motion but it really confirms for me the validity of the vortex theory.

The Omega minus had been predicted by Murray Gell-Mann and its decay in three steps, shedding a dose of Strangeness at each step, lent considerable support to his theory of quarks.[1] However, the decay of the Omega-minus lends as much support to the vortex theory as it does to quark theory. This is an example of how the same experiment can support seemingly contradictory theories.

Something else is revealed by the strange particles. In his Inaugural lecture to the historic chair of Newton at Cambridge Stephen Hawking said that *"...because of the Heisenberg uncertainty principle, the electron could not be at rest in the nucleus of an atom".*[2] The formation of strange particles in atomic nuclei around electron seeds suggests that electrons are to be found lodged within the nucleus of an atom.

Electrons in the atomic nucleus reveal cracks in the Heisenberg cornerstone of quantum mechanics and uncertainty may not be as certain as scientists would like us to believe. But cracks allow in the light. Cracks in the accepted theories provide opportuni-

---

2   **Hawking** Stephen, *Black Holes and Baby Universes*
Bantam 1993

ties for new theories to take hold. Some grip to the cracks, slowly strangling old theories like ivy on dying trees. Others rewrite science, like Darwin's theory of evolution or Einstein's theories of relativity. Like dynamite in the cracks, they blow an entire edifice of assumption apart.

Memories

The immense stability of natural particles against the instability of the manmade particles reveals another crack in the standard theory. The vortex theory suggests natural particles of energy are memories and uses this idea in the quantum laws of motion to account for the unstable particles generated in high energy physics and the property of strangeness associated with them. Memories infer mind and the immense stability of natural particles reflects the stability of the mind that underlies them. The Yogic contention of a universal mind is supported by the vortex theory which in turn stands up to experimental physics. Many in science would choke on this. The alternative for them is the accepted theory of quarks but that is really full of cracks.

Chapter 4

Quark Theory

Will some unknown young scientist find a new way of looking at fundamental physics that clarifies the picture and makes today's questions obsolete?

Murray Gell-Mann

O ver the years people asked me if quarks are quantum vortices. My reply was always no because I had no reason to believe in quarks. I had a simpler way to explain the particle zoo that quark theory was invented to do. When I heard about quark theory in the 1970s, I thought it crude and cumbersome; a resurrection of classic materialism. I joked that if Newton were alive today, he would probably have written to his friend Richard Berkley: *"It seems probable to me that in the big bang, matter formed mainly as quarks."*

On the 27th January 1977, a BBC Television programme, *The Key to the Universe,* reported on the new physics. In a companion book to this programme, under the same title, Nigel Calder wrote about how the theory for quarks originated: *In the early 1930s the contents of the Universe seemed simple. From just three kinds of particles, electrons, protons and neutrons you could make every material object known at the time. Thirty years later human beings were confronted with a bewildering jumble of dozens of heavy, apparently elementary particles, mostly very short lived. They came to light either in the cosmic rays or in experiments with the accelerators. The particles had various mass-energies and differing qualities such as electric charge, spin, lifetimes and so forth...*

A small group of theorists brought order out of chaos. The principle figure amongst them was Murray Gell-Mann of Caltech (The California Institute of Technology), then in his early thirties. He declared that all the heavy particles of nature were made out of three kinds of quarks. He had the word from a phrase of James Joyce 'Three quarks for Muster Mark.' It was the mocking cry of gulls, which Gell-Mann took as referring to quarts of beer, so he pronounced quark to rhyme with 'stork'. Many other physicists rhymed it with 'Mark'. In German, as skeptics were not slow to notice, 'quark' meant cream cheese or nonsense....[1]

There is a lot in a name and the quark theory is nonsense, as the name suggests, not because of mocking gulls but because of a black swan. The philosopher of science, Karl Popper (1902-94) used an analogy of black and white swans to explain that science is not in the business of proving theories but rather of disproving them. Someone could have a theory that all swans are white but even if a thousand white swans were counted, their belief would not be proved true and the addition of more white swans wouldn't make it any truer. However, the appearance of a single black swan would disprove the theory altogether. Scientific theories are white swan belief systems which survive until they are destroyed by the appearance of a black swan, that is, the arrival of even a single fact, which makes it clear that they cannot be true. [2]

The Black Swan

The black swan for the quark theory is the proton. A single fact, which throws the quark theory into question, is the lifespan of a proton. The lifespan of a proton has been estimated at 10^{33} years, that is a billion, trillion, trillion years, whereas one ten billionth of a second is considered a strangely long lifespan for any of the new particles found in high-energy research. As spontaneous proton decays have never been observed, the proton can be treated as being infinitely more stable than any of the new, heavy

1   **Calder** Nigel, Key to the Universe: A Report on the New
     Physics BBC Publications 1977
2   **Popper** Karl, *The Logic of Scientific Discovery* Hutchinson
     1968

particles called *baryons* that have come to light in the accelerators.

There are supposed to be six different types of quark: *up* and *down, strange* and *charm, top* and *bottom*. Theorists speculate that up and down quarks are relatively light and stable whereas the others are heavier and less stable. They account for the difference in stability in terms of the different types of quark. However, the proton is infinitely more stable than even the neutron which is also supposedly formed of up and down quarks. The quark theory doesn't say why there is a difference in stability, why protons stand alone in the particle zoo for longevity; it is just a way of accounting for the difference in lifespan between protons and all the new particles discovered in physics – including neutrons. Quark theory is built around the fact. It does not account for the fact. The fact is protons are stable whereas all other heavy parti-

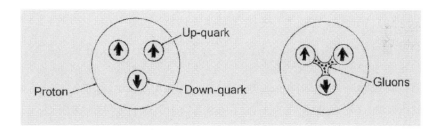

cles, including mesons and neutrons, are not and quark theory does not explain why whereas the vortex theory explains clearly and simply why it is so.

Imagine walking down a road between two building sites. On the site to the left, the houses disintegrate as soon as they are built. Within minutes of the last lick of paint being applied, they are gone. On the site to the right, the houses are advertised for sale with a billion, trillion, trillion year guarantee. It would be crazy to assume that similar bricks, mortar and construction techniques are employed on both building sites without providing an account for the difference in the durability of the houses. Yet in physics today physicists claim protons and all other baryons including neutrons are constructed of quarks cemented together by gluon bonds.

Quark theory could be used to explain infinitely stable protons or it could be used to account for the unstable new particles but not both! To use quark theory to explain only protons would be pointless, because the theory was invented to explain neutrons and the new particles discovered in high energy physics. But if quark theory provided an explanation for all the baryons apart from the proton, it would still be pointless because the great bulk of the mass of the physical Universe consists of protons, whereas the other baryons – apart from the neutron – have been observed only in high-energy experiments. If the theory were to exclude protons the worldwide programme of high-energy research, dominating physics, would be meaningless.

There is no need for quark theory in the real world as the zoo of short-lived particles have been synthesized out of the massive amounts of energy fed into the high-energy particle accelerators. They have no place in normal matter. They don't normally exist! They are just products of high-energy research, anomalies of dubious value created in enormously expensive experiments.

When the top-quark was supposedly discovered at Fermilab in April 1994, Jim Dawson opened his report in the Minneapolis 'Star Tribune' with the statement: *"So there we have it, after more than two thousand years of searching, all of the fundamental stuff of Democritus' atom has been revealed. The crowning moment came a couple of weeks ago, when physicists announced that a gigantic, 5,000-ton machine apparently had detected a very small particle called the top-quark."*[3]

3   **Dawson** Jim, *Star Tribune of Minneapolis*, May 15th 1994

Quark theory is based on the material delusion.

The physicists didn't actually see a top quark. All they saw was the tracks of a jet of electrons and muons which they supposed to be the breakdown products of W-particles which they took to be the remnants of a top-quark. The recent discovery of Higgs Boson at CERN was an interpretation of similar fallout. While everything in the world is supposed to be made of quarks, not a single quark has ever been observed in a free state.

Another problem with the quark theory is a quark in a proton was calculated to have five times the mass of the proton itself. Physicists responded with a convenient suggestion. When quarks form a proton they lose fourteen fifteenths of their mass. As equations show nuclear binding increases with loss of mass, they contended that the quarks would then be so tightly bound that a quark could never be free. That is why no one has ever seen a quark!

Some physicists are unhappy about the discrepancy between the mass of the quark and the mass of the proton. Richard Feynman is on record saying *"The problem of particle masses has been swept in the corner."*[4]

Quark theory became generally accepted in 1968 after an experiment at the Stanford Linear Accelerator in California (SLAC). In the SLAC experiments, electrons were accelerated down a three-kilometer long 'vacuum tube' by intense radio pulses and then targeted on protons in liquid hydrogen. The results of these experiments showed that electrons were being scattered, or bounced back from what appeared to be something small and hard within the protons in the hydrogen atoms. From this it was inferred that the protons were not truly fundamental but contained smaller particles.

4   **Calder** Nigel, Key to the Universe: A Report on the New
     Physics BBC Publications 1977

Physicists were looking for quarks and naturally concluded that their bombarding particles were bouncing off quarks in the proton so these experiments caused the quark theory to become ac-

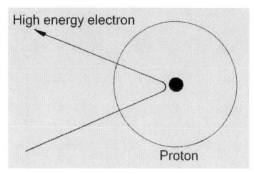

cepted into mainstream physics and led to a Nobel Prize for Gell-Mann in 1969.

The SALC experiments were supported by experiments at CERN, the European Nuclear Laboratory near Geneva. At CERN, protons were accelerated in vast intersecting rings. At the intersections, energetic protons were directed into head-on collisions. In the high-energy impacts, new particles were formed and came out at right angles to the beams. Again these results seemed to confirm the quark model. However, the SLAC and CERN experiments inferred rather than proved the quark model. They merely demonstrated that the proton has 'a hard heart'.

Thinking about the vortex theory while he was digging his garden, my father in law, John Bingham, came up with a different explanation for the hard hearted proton. He suggested as a vortex the proton could be treated as a system of increasingly compact lines of force. I was able to use his idea to explain SLAC because I realised bombarding particles would compress these lines of force, and being quantum vortices themselves, they would undergo reciprocal compression. The compression of the increasingly compact energy toward the vortex centres could result in hard impacts between the colliding particles which could cause the scattering observed at SLAC and CERN. My own father said physicists were obsessed with 'atom bashing'. He said nature was subtle; we should seduce her for her secrets, not beat them out of her.

During the Second World War physicists succeeded in harnessing the energy locked in the atom to make the most terrible weapons of destruction ever known to mankind. Governments sprinting in the arms race rewarded them with virtually unlimited budgets for continued nuclear research. Vast sums of money spent on building, maintaining and running particle accelerators enabled scientists to create a host of new particles. The need to account for these led to a 'nonsense theory'. The theory threw up new difficulties and exciting predictions, which then required more research and more powerful accelerators and of course a lot more money.

Intent 'to throw good billions after bad', the Large Hadron Collider has been built at CERN in Geneva. Costing billions of euros, it is the biggest accelerator in the world. This monster machine is twenty times more powerful than any built previously. Its main purpose was to identify the source of mass – now believed to be the Higgs Boson – and to simulate the first moments of the Big Bang. It has also been built to further research into quarks and force carrying particles that probably don't exist. Creating a new generation of particles, this euro-gobbling accelerator will lead to more elaborations on the theories, which would require more research and eventually the need for an even bigger accelerator.

High-energy research has created an endless cycle in which research physicists create problems for theoretical physicists to solve and they, in turn, predict yet more problems to research. While this elaborate game ensures indefinite employment for scientists and their army of technicians, it produces information which is of little relevance or value to the unfortunate tax payers who have to foot the enormous bill.

Flying Pig Research

That reminds me of a sad world of sick, homeless and starving people where there lived a mad professor who believed in flying pigs. No one had ever seen a flying pig but the mad professor managed to convince all the universities in his world that hunting for flying pigs was the most valuable line of research they could undertake. As the greatest minds in the nutty world caught flying pig fever, the developed nations fell over themselves in the frenzy to build bigger and ever more elaborate flying pig blasters. No expense was spared. Money taken from the people that could have been used to improve the quality of their lives was squandered in the search, but after decades of hunting no one spotted a single flying pig. After all that effort you would have thought that the learned professors would have accepted that their flying pig theory was cranky, but not willing to admit to being cranks, they just went on building ever bigger and more powerful blasters.

Particle accelerators have been of value, that is undeniable, but the question is whether the expense of building ever bigger ones is justified. Considering the need for cutbacks in public spending and the law of diminishing returns perhaps it is time for a day of reckoning at CERN, especially in view of something else that has been going on that really needs to be accounted for:

Plum Pudding Theory

Another nut case, in the crazy world, was a mad professor of puddings! A *student* in his department of nutrition baked plums in a pudding then ran round shouting, "Eureka, I have just discovered plum pudding!"

The professor was not impressed. He snorted over his spectacles, "That is not a plum pudding you have baked, it's a Black Forest gateau."

"But," argued the downcast student, "I mixed plums into my pudding before I baked it and when I weighed it, the weight was that of the plums and the pud and at the end, when I shook the pudding, out came a plum. It has to be a plum pudding!"

The professor became really angry. "You stupid student, have you learnt nothing of what I have taught you? When you bake plums in a pudding due to the interaction of a weak cooking force, the plums change flavour into cherries and the pudding becomes a Black Forest gateau."

"Well how do you explain the plum that fell out of the pudding?" dared the student; "There are no plums in a Black Forest gateau!"

"The gateau is unstable," retorted the professor in a fury, "after a few minutes it falls apart reverting to the original plums and pudding. It reverse flipped its flavour silly boy!"

How could the student argue? He was speaking to none other than the President of the Royal Society of Puddings. If he wanted to make it in his mad world and get a good career in the cake and pudding industry, he had to accept that a pudding baked with plums transformed into a Black Forest gateau...

Light electrons with a negative charge are plums in the satire. Protons, with a positive charge, are 1836 times as massive as electrons. They are the pudding mass. The plum pudding is a neutron; the third particle in the atom.

The neutron is neutral in charge and can be formed out of an electron and proton. That happens in a process called K-capture. It is the sum mass of electron and proton and in Beta decay a neutron can fall apart into an electron and a proton.

From those facts it is pretty obvious a neutron is an electron bound to a proton. However, don't suggest that to a professor of physics. He might get very annoyed and say,

"Don't jump to hasty conclusions! You should know neutrons are not electrons bound to protons. Electrons cause protons to change quark flavour due to the interaction of the weak nuclear force. That is how they come together to form a neutron. This then reverses when the unstable neutron decays."

If you accepted his authority on matters subatomic he might detail that the proton is made up of two up-quarks and one down-quark, explain that the neutron is made up of two down-quarks and one up-quark and go on to say that when a proton interacts with an electron the weak nuclear force comes into play which transforms the flavour of an up-quark into a down-quark and causes the electron and proton to vanish, their place being taken by the neutron.

He would then explain as the neutron is unstable they can come back into existence which would occur if the process is reversed with the escape of a neutrino.

If you get lost or bored and want to make an escape say you discovered quark. If he asks when you were last at CERN reply it wasn't there you found it, it was in the supermarket between the yoghurt and the cream cheese

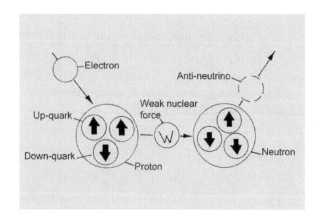

The Lick Experiment

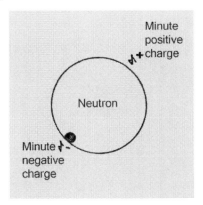

Meanwhile... back in the nutrition department of the crazy world, all was not going well with the mad professor of puddings. Another student baked plums in a pudding then licked it. He tasted a plum and so established that the plums did not lose their flavour when baked in a pudding. The mad professor was very embarrassed by the arrival of that awkward fact. He ranted and raved and made it a rule that food was never again to be tasted in the lab. And then he used his prestige to ensure that when the result of the lick-experiment was printed in the textbooks, it was set in small type so that hopefully, it would be overlooked as an unimportant fact....

The 'lick-experiment' parodied an experiment in 1957 in which the neutron was discovered to have a slight electric charge. This means it is not entirely neutral. On one spot it displays a minute negative charge, in the order of a billion, trillion times weaker than that of a single electron.

The measure is considered too weak to be taken as conclusive but it did suggest the neutron was a bound state of opposite charges, which mostly cancel each other out. [5]

The presence of charged particles in the neutron is supported by the fact it has magnetism i.e. it has a measured magnetic moment[5]. The neutron, if it were a truly neutral particle, would have no magnetism because the magnetism of a subatomic particle is created by the spin of its charge.

Another problem with quark theory is the totally arbitrary assumption that quarks have fractional charge. Up-quarks are supposed to have 2/3 charge and down-quarks 1/3 charge. In the proton 2/3 + 2/3 - 1/3 = 1 which gives the proton unitary charge. In the neutron 1/3 + 1/3 - 2/3 = 0 gives the neutron zero charge. However, there is no evidence for fractional charge in nature. This is a speculation heaped on speculation.

I was inspired in my work by the final paragraph in my university physics textbook: *"Let us hope that out of this chaotic riddle will come a profound and simplifying answer. We may be likened to those who knew only Ptolemy's complex description of the solar system. What we need is a Copernicus to assimilate and interpret the data with a generalisation which will not only solve the riddle, but lift our sights to levels we cannot now foresee."*[6]

Why did physicists argue against the obvious structure of the neutron? Why did they concoct an elaborate theory of up-quarks changing to down-quarks in an alchemy of particle transmutation? There has to be a very good reason why they were so determined that a neutron was not an electron bound to a proton.

5   **Segrè** Emilio, *Nuclei & Particles* Benjamin Inc 1964
6   **Richards**, et al, *Modern University Physics*
     Addison-Wesley 1973

Were physicists trying to hide something; some really awkward fact?

I believe the overwhelming evidence of a neutron being an electron bound to a proton is rejected by physicists because it threatens their most cherished theory;' quantum mechanics.

Chapter 5

Quantum Mechanics

I think I can safely say nobody understands quantum mechanics
Richard Feynman

The Uncertainty Principle

Quantum mechanics depends on a principle proposed in 1927 by a German physicist Werner Heisenberg (1901-76). Heisenberg suggested that the process of making certain measurements in the sub-atomic world would increase the uncertainty about what was going on there. For example, if you wanted to look at a particle in order to ascertain its position, you would reflect light off it. But the act of bouncing light off the particle would give it a kick that would increase its momentum and so make its position more uncertain.

Looking at small objects requires more energy than is required for looking at large ones. This is evident in the electron microscope, which employs higher frequency radiation than a normal light microscope. Because sub-atomic particles are so small, the action of ascertaining their position would give them such a kick you would never know where they are. Heisenberg argued against being able to determine the position and the momentum of particles. His principle of indeterminacy is known as the Uncertainty Principle. Albert Einstein despised Heisenberg's principle. He described it as, *"A real witches calculus...most ingenious, and adequately protected by its great complexity against being proved wrong."*[1]

1   **Matthews** R *Unraveling the Mind of God* Virgin 1992

At a Solvay Conferences in the 1920's Einstein disputed with Heisenberg through the night reducing the younger man to tears. Imagine arguing all night with Einstein; that would be enough to make anyone cry!

The Heisenberg uncertainty principle was never an easy law to test experimentally because that required a certain measure of the simultaneous position and momentum of a sub-atomic particle, which the principle stated was virtually impossible to obtain with certainty. However in 1932 the neutron was discovered and it offered way to test the principle.

If a neutron is treated as an electron combined with a proton, the position and momentum of the electron could be determined with a high degree of certainty. However, when the witches' calculus was applied to neutrons it predicted that electrons bound to protons could have velocities up to 99.97% of the velocity of light. Imagine electrons in neutrons rushing round at the speed of light; that would be mad!

One way to save the uncertainty principle would be to say it does not apply to neutrons. Electrons in the atom are already excluded. If physicists make too many exceptions to protect the principle they could be accused of rejecting experiments that challenge the validity of their theories and as Richard Feynman said, *"If your theories and mathematics do not match up to the experiments then they are wrong."*[2]

The uncertainty principle is a major stumbling block to unifying Einstein's relativity and quantum theory. Stephen Hawking commented, *"The main difficulty in finding a theory that unifies gravity with the other forces is that general relativity is a classical the-*

---

2   **Calder** Nigel, *Key to the Universe:* BBC Publications 1977

ory in that it does not incorporate the uncertainty principle of quantum mechanics."[3]

That was a real slight on Einstein as it was he who broke the mould of classical physics by describing matter as a form of energy. The fact is the uncertainty principle failed its 'neutron test' a decade after Einstein had been elbowed out of quantum physics. He knew the principle was wrong.

Many people will protest that Heisenberg's principle must be right because of its incredible success in quantum physics. However, the successful application of a principle does not prove it is valid. Just as a car can work without a road worthy certificate a theoretical principle can work even if it is wrong. With the failure of the uncertainty principle and the quark theory it could be said that the Standard Theory is running without a certificate of road-worthiness. Maybe the time has come for it to be scrapped!

Quarks

Rather than admit to the uncertainty principle being unsound, physicists have done their best to dismiss the awkward facts about the neutron and concocted the complex quark account for it instead. Like the mad professor of puddings they deny that the neutron is an electron-proton pudding and insist it is a quark gateau instead! It is hard to see how they could do otherwise because if Heisenberg's principle was shown to be invalid, then the entire edifice of quantum mechanics would collapse and to the modern physicist that is unthinkable. In the words of A.J. Leggett: *"Quantum mechanics...has had a success which is almost impossible to exaggerate. It is the basis of just about everything we claim to understand in atomic and sub-atomic physics, most things in condensed-matter physics, and to an increasing extent much of cosmology.*

3   **Hawking** Stephen, *Black Holes and Baby Universes*
      Bantam 1993

Chapter 5: Quantum Mechanics

*For the majority of practicing physicists today it is the correct descrip-
tion of nature, and they find it difficult to conceive that any current or
future problem of physics will be solved in other than quantum mechanical terms. Yet despite all the successes, there is a persistent and, to their
colleagues, sometimes irritating minority who feel that as a complete
theory of the Universe, quantum mechanics has feet of clay, indeed 'carries within it the seeds of its own destruction'."* [4]

Quantum mechanics and the uncertainty principle moved centre
stage in physics with the idea that forces could be carried between
particles by the exchange of other particles. It was suggested that,
within the bounds of sub-atomic uncertainty, particles could bor-

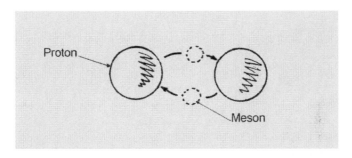

row energy from the Universe to bring about the creation of
short-lived force-carrying particles. So long as these 'virtual'
force-carrying particles were sufficiently unstable to decay and re-
pay the energy debt within the time allotted by Heisenberg's formula then no conservation law would have been broken in the
overall process of their formation and demise. If the time span of
their existence was short enough, the uncertainty principle allowed
for very large amounts of energy to be involved in the formation of
the particles, which would, in turn, enable them to carry very powerful forces.

4   **Leggett** A.J., *The Problems of Physics*, Oxford Uni. Press
     1987

The underlying premise that 'anything is possible behind the screen of uncertainty' was a license to speculate. The idea, that sub-atomic particles run credit with the Universe, as though it were a bank, is crazy but because it is impossible to be certain that it doesn't happen, physicists were getting away with saying it does happen. Quantum mechanics would seem to confirm the law of projection: 'We see the world, not the way it is, but the way we are'.

With his mortgage, bank loan, and credit cards, the professional physicist has projected the image of credit on the sub-atomic world. Mr Proton could be imagined to borrow energy in order to create gluons for the binding of quarks, W and Z particles for the weak nuclear force, mesons for the strong nuclear force, vir-

Mr Proton

tual-photons for the forces of electric charge and magnetism, gravitons for the force of gravity and now Higgs Bosons to account for its mass. Attempts are being made to consolidate the borrowing from the universal energy bank into one grand unified 'GUT' account. So far this has met with little success.

Hideki Yukawa

In 1934, a Japanese physicist, Hideki Yukawa (1907-81) used the uncertainty principle to predict the existence of meson particles to carry the strong nuclear force. Doubts about the energy credit system of quantum mechanics dissolved when in 1947 Cecil Powell discovered mesons that fitted Yukawa's prediction. Mesons fitted the math of the witches calculus and so Yukawa and Powell shared the Nobel Prize. While it cannot be denied that mesons exist, just because Yukawa predicted them does not mean that the quantum mechanical explanation for them is correct because I was able to use the same experiment to support an entirely different theory. Predictions drive science, suggesting directions for research and designs for experiments. The problem comes when they lead to speculations that primary particles have a credit rating! Increasingly crazy theories have become thenorm in physics. This prompted Niels Bohr to jest at the end of a lecture given by Wolfgang Pauli, in 1958, *"We are all agreed that your theory is crazy. The question which divides us is whether it iscrazy enough."*[5]

Quantum Electro Dynamics

People demand proof for truth. They want to believe ideas can be proven by a scientific experiment but science is not in the business of establishing truth. Predictions establish the use, not the truth of a theory. Many of the most successful theories in science defy common sense. An example of this was Richard Feynman's Quantum Electro Dynamics. QED is a very successful theory. It is a branch of quantum mechanics with a margin of error in the order of one part in three billion. However, it requires us to believe protons and electrons borrow energy to make virtual photons of light to account for their charge. QED depends on the uncertainty principle.

If the uncerftainty priciple was invalidated by the neutron, QED could collapse. Should that happened the scientific method would be compromised and scientific criteria for truth could be challenged. People might begin to argue that scientists are simply proving what they want to believe. Many people these days believe that we create our own reality. They could shout *"placebo!"* contending that experiments reflect the beliefs of the scientists performing them.

Now maybe you can see why mention of the neutron as a bound state of electron and proton upsets so many physicists? Quan-

5   **Calder** Nigel, *Key to the Universe:* BBC Publications 1977

tum mechanics cannot stand if the neutron invalidates the Heisenberg uncertainty principle. If the neutron is a bound state of electron and proton then the outstanding success of quantum mechanics will have done little more than reveal just how unreliable science can be. It is hardly surprising that when the neutron appeared as a black swan over the quaint little pond of quantum mechanics, instead of welcoming its arrival – as scientific integrity would demand – physicists tried to shoot it. Had they allowed it to land, Heisenberg's principle would have been a dead duck and virtual force-carrying particles, its darling little ducklings, would have certainly drowned. Instead they called out 'quark quark', muddied the waters with incomprehensible math, and shooed it away.

Emperor's new Clothing

It is the story of the Emperor's New Clothing all over again. Quarks and the virtual particles of quantum mechanics are phantoms but people in physics won't admit to it because there are too many jobs and academic reputations at stake. However, there is no future for science if awkward facts are swept under the carpet to save the theories. In the words of George Gamow, *"Staggering contradictions of this kind, between theoretical expectations on the one side and observational facts or even common sense on the other are the main factors in the development of science."* [6]

In his Inaugural Lecture entitled 'The End is in Sight for Theoretical Physics', Stephen Hawking said that because of the Heisenberg uncertainty principle, the electron could not be at rest in the nucleus of an atom. The fact is that electrons are at rest

---

6   **Gamow** George, Thirty Years that Shook Physics,
      Heinemann

in the nucleus of an atom and because of that the end is in sight for theoretical physics.[7]

It seems like divine retribution that the particle used by physicists to detonate weapons of mass destruction should threaten to overturn their most cherished theory. However, this may not be a bad thing. Appearing as harbingers of change, the evidence provided by the natural particles could clear out falsehoods in physics and prepare the way for a new understanding of the Universe.

There is a positive side to Heisenberg's theory. If through the uncertainty principle Heisenberg has exposed uncertainty in the scientific criteria for truth, he will have released humanity from the burden of truth imposed by science. Religions have a history of telling people what they can and cannot believe. Since the *enlightenment* there has been a tendency for secularists to impose their understanding of the Universe on people in a way that is engendering an attitude of intolerance, unbecoming of science.

If the neutron has unhinged the Heisenberg uncertainty principle, no scientist can claim to be certain of the truth. As waves breaking endlessly on the shore so future generations will throw their interpretations of truth on a beach of uncertainty. No longer should anyone be able to say for certain, 'this is the way it is, I am right and you are wrong'.

Heisenberg was a great and brilliant scientist. His ideas dominated physics in the 20th century. His lasting legacy could be his contribution to the infinity of possibility. He has helped in the emancipation of humanity, freeing us from the burden of scientific and religious truth so that every human spirit can be at liberty to believe whatever he or she chooses. Through the

7   **Hawking Stephen** *Black Holes and Baby Universes*
      Bantam 1993

philosophy of uncertainty Heisenberg may have done more for the truth than any saint or scientist because the paradox is, the less certain we are of truth the closer we are to it.

Chapter 6

Quantum Theory

A scientific truth does not triumph by convincing its opponents and making them see the light, but because its opponents eventually die and a new generation grows up that is familiar with it.

Max Planck

In physics the *quantum* is the minimum energy involved in an interaction. Max Planck (1858-1947) suggested heat and light are radiated in packets for which he used the term quantum originally coined by Helmholtz. Einstein then suggested that a quantum of energy maintains its particle nature throughout its life. In 1913 Niels Bohr (1885-1962) applied the quantum idea to electrons in atoms. He argued that the electron could only gain or lose energy in discrete quantum bundles, which would shift it from one orbit to another.

As Bohr applied the quantum principles of light to particles of matter Louis de Broglie (1892-1987) proposed wave-particle duality when he suggested that the classical distinctions between matter and light be dropped altogether and everything be treated as waves. Erwin Schrödinger (1887-1961) supported him by applying his wave equation to the electron orbits and so in quantum theory the distinction between matter and light began to dissolve. Schrödinger also showed that the electron orbit could not be treated as certain. The position of the electron in the atom could only be deduced in terms of where it would most probably be according to its energy.

Einstein was deeply unhappy with the improbable direction of quantum theory. According to the renowned philosopher of science, Thomas Kuhn (1922-96), more credit goes to Einstein than Planck for the foundation of quantum theory[1]. Yet the bright young men of the quantum revolution chose to ignore the warnings of the greatest mind in science.

Uncertainty crept into quantum theory when in 1925, the twenty four year old Werner Heisenberg, joined the team. Einstein despised the uncertainty principle and argued against the younger man's ideas but Niels Bohr, while retaining his friendship with Einstein, sided with Heisenberg. Bohr came to dominate quantum physics because of his charisma and popularity. Einstein was more of a recluse. Bohr was an unpretentious, energetic, much loved man who was always concerned for the welfare of others. Through his powerful intellect and success as a fundraiser and administrator he made his Copenhagen Institute the world centre of theoretical physics. His outstanding human qualities drew the younger men to him like a magnet. But some were unhappy about Bohr's dominating influence over the burgeoning quantum theory. As Robert Matthews put it in 'Unravelling the Mind of God': *"The Copenhagen interpretation of quantum theory remains the most widely accepted way of looking at quantum theory amongst scientists today, but it should be stressed that, despite what some might claim, it remains just that: an interpretation. There is powerful experimental evidence that it is an acceptable way of looking at the world, but it is most definitely not the only possible interpretation"* [2]

Despite the outstanding human qualities of Bohr, Robert Matthews described Bohr's influence on quantum theory as the 'quantum bandwagon'. In his book he continued: "At the reins of the quantum bandwagon were one or two superstars - like Niels Bohr – and in the back were like minded theorists and

1   **Kuhn,Thomas**, *Black-Body Theory and the Quantum Discontinuity: 1894-1912* Clarendon Press, Oxford, 1978
2   **Matthews** R *Unraveling the Mind of God* Virgin 1992

experimentalists armed to the teeth with laboratory results to prove their case. If you get in the way of such a bandwagon, you will at best be left behind, and at worst be flattened."[3]

Matthews then quoted David Bohm (1917-92) as saying: "Most physicists said they followed Bohr without really knowing what Bohr was doing."

1925 to 1928 were the years of 'Knabenphysik', 'the boy physics'. Radical new ideas for quantum mechanics were emerging from the minds of Werner Heisenberg, Wolfgang Pauli, Max Born and Paul Dirac. Pauli was famous for his exclusion principle and Einstein was excluded.

The older generation including Bohr, Einstein and Schrödinger struggled to keep up with the bright boys. Bohr, revered by the younger men, enjoyed their friendship and was indefatigable in clarifying the detail of their emerging theory of quantum mechanics. It is no wonder the new generation of physicists put more store by Bohr than Einstein.

Sadly the greatest scientific genius in history was swept aside in the storm pouring out of young, tempestuous heads. In a spoof on 'Faustus' performed at the Copenhagen Institute of Physics in 1932, Paul Dirac ended the play with: "*Old age is a cold fever that every physicist suffers with, when he is past thirty he is as good as dead.*" [4]

In the eyes of the emerging physicists Einstein was a 'has been'. He refused to acknowledge that at root the world was essentially and ineluctably unknowable. He pioneered quantum theory and led it for many years then became its most redoubtable critic. It is easy these days to overlook the fact that quantum theory origi-

3   **Matthews** R *Unraveling the Mind of God* Virgin 1992
4   **Segrè** Gino, Faust in Copenhagen: A Struggle for the Soul of Physics, Viking (2007)

nated from Planck and Einstein not Heisenberg and Bohr. People tend to forget Einstein was the leader of the quantum revolution which upturned classical physics; the brightest star in the original quantum firmament.

The Philosophy of Realism

In their version of quantum theory, Bohr and Heisenberg drew the distinction between classical and quantum physics on the issue of realism. Up until the turn of the 20th Century, there was a tendency to believe there was an ultimate reality that man, through experiment, could discover. Scientists believed that they could explain, in their theories, what these realities were. This way of thinking came to be known as the *philosophy of realism*. As scientists had to adopt different and incompatible, wave and particle models to explain the same phenomenon, the inadequacy of realistic models became apparent.

This led to the Copenhagen interpretation, which rejected realism as naïve and denied that the theoretical models could ever be true representations of reality. While this was true, the great richness of classical physics lay in its simple models, such as lines of force and vortex atoms which made the subject understandable. The problem was, as realistic models were abandoned unrealistic representations of reality took their place.

Physicists have not been entirely happy with the Copenhagen approach to quantum reality. Heinz Pagels said, *"Something inside us doesn't want to understand quantum reality. Intellectually we accept it because it is mathematically consistent and agrees brilliantly with experiment. And yet the mind is not able to rest."* [5]

Throughout the 20th century physicists failed to maintain a consistent posture. Like most people, they tended to think of elementary particles as 'tiny little things' with definite properties

5   **Pagels** Heinz, The Quantum Code, Michael Joseph 1982

such as mass, spin and electric charge. They could make beams of them and bash them together like billiard balls. Whilst quantum idealists proclaimed that naïve realism was a thing of the past, most physicists were behaving as though they had a deep conviction that their hypothetical particles were not just models but that they really did exist. For example, with their particle accelerators, physicists hunt for quarks at the core of matter as though they actually exist!

Quantum theory was established by Planck and Einstein, on the principle that everything is formed out of particles of energy and the distinction between classical and quantum physics should be hinged on this epoch making principle, not the red herring of realism. Planck broke ranks with classical physics by suggesting radiant energy is particulate. When Einstein proved this he also made it clear that matter as well as light consists of particles of energy not particles of material substance. He established we live in a non material world and materialism is naïve realism!

Professionalism

The story of quantum theory reflects personalities and the extent to which these influence the direction of physics. Professionalism can also be a threat to physics. Professional physicists were jealous of Albert Einstein because he was an amateur. They were determined to topple him from the pinnacle of physics.

Career physicists have least to gain and most to lose from fundamental changes in physics. Professionals are trained to maintain and teach the status quo. Paradigm shifts can threaten careers and incomes. Professionals rarely welcome new directions that undermine everything they have been taught and everything they teach; especially if they come from non professionals! They are more likely to ensure reliable operation of Russell's Law that: *The resistance to a new idea is proportional to the square of its importance.*

I am not a professional physicist. I have written my theory so non scientists can understand it because it is only people without a vested interest in something who will have the impartiality to accept changes to it. More and more people these days are realising the importance of standing in their own power and deciding for themselves rather than rely solely on professional opinion. The purpose of education is not to tell us what to believe but to equip us with the tools so we are in an informed position to decide for ourselves what we want to believe. As much as anything the vortex theory is intended to foster original and independent thinking.

Planck knew progress with a new theory is not made by converting people of an established mind set. He recognised his quantum theory was more likely to spread amongst younger minds that were open to new ideas. His attitude was to wait for those set in their ways and habits of thinking to die off so a new generation, familiar with his theory would carry it forth; which is precisely what happened. A new approach to quantum theory emerges from the vortex theory that is simpler and easier to understand than quantum mechanics and should appeal to a new generation.

Chapter 7

Mechanics of the Quantum Vortex

These fundamental things have got to be simple

Lord Rutherford

Quantum theory is about particles of energy and quantum mechanics determines how these particles interact. My contribution was to bring the vortex from Yogic philosophy into quantum theory to describe how particles of energy form matter. The mechanics of the quantum vortex explains how wave particles of energy interact with vortices.

The Egyptian philosopher Hermes, famed for the fractal principle – *As above so below: As below so above* – is purported to have taught that everything is motion and that male-female duality occurs at every level in the Universe.

Light could be described as a 'masculine form' of energy because particles of light are akin to sperm with their wave-propagation form of motion.

The word matter is derived from the Latin, *mater* for mother and matter could be treated as the 'feminine form' of energy because the subatomic vortex appears much like the passive, receptive, spherical ovum.

Sexual symmetry appears in the interactions between wave trains and vortices of energy. Just as the active sperm drives into the passive ovum, so the wave train of energy appears to drive into the vortex. Just as the ovum is receptive to sperm so the vortex particle seems to be receptive to the wave particle of energy.

The quantum vortex would be receptive to wave-train particles of energy if it is treated as a spiral path of energy. Just as a car follows the contours of the road so, in this theory, the propagating energy would be imagined to follow the spiral contour of the vortex into its centre.

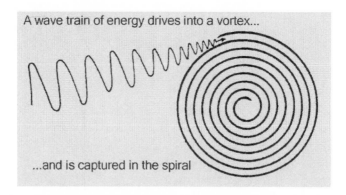

A wave train of energy drives into a vortex...

...and is captured in the spiral

The flow of energy is unidirectional. We experience this as the flow of time from the past through the present into the future. Just as time doesn't reverse, once a wave train of energy drives into a quantum vortex it can't reverse out. It is effectively captured by its forward motion into the vortex.

Partial Capture

The quantum laws of motion are as much concerned with the propulsion of mass into wave-kinetic motion as with the transformation of wave-kinetic motion into mass. A key idea is that particles of energy can *share inertias*. The concept is that wave energy driving into the vortex could be partially impressed into the shape of the vortex; partially transformed into mass. The vortex moving under the impact could then take on the wave-kinetic inertia of that part of the quantum wave train not transformed into mass. This explains the electron vortex with its low mass and low passive inertia taking on wave kinetic motion; it has assumed the inertia of the wave train of energy. I used this idea of the *partial capture* of a quantum of kinetic energy by the vortex to account for the wave-particle duality of subatomic particles.

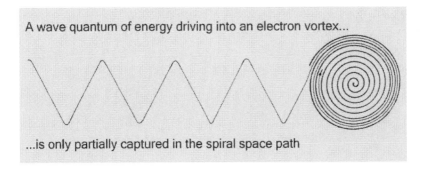

A wave quantum of energy driving into an electron vortex...

...is only partially captured in the spiral space path

The partial capture of kinetic energy by the vortex is illustrated by a tadpole. The head of the tadpole represents the spherical vortex and its tail, the partially captured wave train of energy. A tadpole swims in wave motion because it is propelled by the wave motion of its tail, so the electron vortex would move in wave motion because it is propelled by a partially captured wave-train quantum of energy.

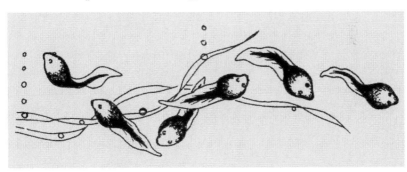

In the pond, a few tadpoles grow massive by devouring their lesser siblings. This image of the cannibal tadpole depicts an elementary particle growing more massive as it is accelerated toward the speed of light. As additional wave trains of energy drive into an electron vortex they cause it to accelerate but its mass would also grow due to partial transformation of the additional energy into mass. The ever increasing inertial mass of the electron vortex would require ever more energy to accelerate it through additional increments of velocity.

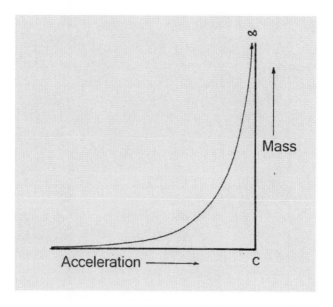

It is obvious that a vortex particle of matter could never move faster than the speed of the propelling wave trains of energy driving it forward. Growing ever more massive, its increased vortex inertia would give it capacity to capture ever more energy. This explains the exponential increase in the mass of particles of matter with acceleration, predicted by Albert Einstein and confirmed in particle accelerators, and why electrons accelerated toward the speed of light have greater mass than electrons at rest.

The sharing of inertias occurs when the passive of the vortex is insufficient to resist the kinetic of a quantum driving into it. Instead of staying still it moves.

The Monkey and the Nuts

The way wave and vortex particles of energy share there s is illustrated by a monkey and a jar of nuts. The monkey represents the wave train of energy and the jar of nuts, the vortex. The monkey can get its hand into the jar, but with a fistful of nuts it can't withdraw so it caught by the jar. The jar is passive in the process, which illustrates the passive role of the vortex in energy capture – it is a spiral path to the quantum wave-train that takes the ac-

tion of driving into it. Rather than let go of the nuts the monkey runs off with the jar which has effectively captured the monkey.

The jar takes on the mobility of the monkey. That depicts the vortex taking on the active inertia of the kinetic energy it has captured. The jar is heavier because it contains the hand of the monkey as well as the nuts. The hand in the jar represents the part of the wave train that has taken on the passive of the vortex; which has increased its mass.

The vortex presents an every tighter spiral path toward its centre so the first wave-train of energy it captures would be tightly bound. Consequently, the resultant bound state of wave and vortex would be very stable. This accounts for the perpetual wave energy of the ground state electron in an atom; why it orbits indefinitely; why atoms are so durable.

Marriage and Sharing

The vortex version of quantum mechanics reveals the symmetry between human coupling and the way vortex-wave couples behave: *As above so below, as below so above.*

Traditionally women didn't propose. Attractive to a man, it was the man who took the plunge. So it is in the quantum world, the vortex is passive but irresistible to a wave-train because of its receptive spiral path. The secure union of vortex and first quantum wave train is represented by marriage between man and woman. This corresponds to the ground state electron in its atomic home. In a marriage each partner imposes their personality on the other. So it is with the quantum couple; each form of energy imposes its inertia, on the other. A harmonious marriage is all about sharing. A successful marriage represents the sharing of inertias between a wave-particle quantum and a vortex. It depicts the *shared state* of kinetic and static inertias.

Men without women can be dissipated or they may gang up to plunder and destroy as a gang of hooligans or an army on the rampage. Women without men can be frustrated and traditionally they couldn't be mothers. Without kinetic energy matter drops to absolute zero. Without matter heat dissipates. Nothing useful happens when the masculine and feminine forms of energy, the static and kinetic forms of inertia, are not in the shared state. This tendency to dissipation and increased disorder is *entropy*.

The opposite of entropy is work. When men are coupled with women rather than go to war they go to work. Useful processes in physics, resulting from the intercourse of kinetic energy with matter, are called *work*. The wave-vortex *couple* leads to useful structures; first atoms then molecules and then the living cells from which we are formed.

Wives can interact with other men as well as their husbands and it is not uncommon for a wife to accept the advances of another man. She may become excited but the ensuing adventure doesn't last because the triangle between husband, wife and lover is unstable. Likewise an electron in an atom can have a liaison with a quantum of energy and take a *quantum leap* into an excited-state. But the excited state is unstable. It leads to change and often doesn't last.

Sometimes the electron will drop from the excited state back to the ground state again. The quantum like a disillusioned lover then leaves as a photon of light. This is photon emission and it accounts for the light emitted by a flame. The yellow colour of a candle flame represents the frequency difference between the ground state and excited state in carbon atoms and the flame represents the temperature zone in which quanta have sufficient energy to excite electrons in hot air currents to take a quantum leap.

The continual stream of light emitted by a flame or an electric light is generated by the vibration of electrons between the ground state and the excited state; by electrons accepting and ditching quantum lovers many times a second. Strike a match and you will see a constant stream of old flames leaving the burn! *As below so above, as above so below.*

The excited state is unstable because the subsequent quantum of energy corresponding to the lover is not captured in the tight central spiral of the electron vortex because that spiral path is already occupied by the original quantum corresponding to the husband. This weaker bond with the electron vortex corresponds to the weaker social bond between a wife and lover than a wife and husband.

Chemistry

A wife may run off with her lover and settle in new homes. She may not return to the old home or may spend some time in the original home and some time in the new. This happens in quantum society. It is called chemistry!

Just as the chemistry has to be right for a woman to accept a man so it is between atoms for chemistry to occur! A quantum must have the energy, the minimum frequency required to excite an electron sufficiently to leap into a higher excited energy state in the atom. Einstein won the Nobel Prize for showing a quantum had to have a minimum frequency to carry an electron away from its atomic orbit.

In the candle flame, not all electrons revert to the ground state in their atoms. Once excited, some leap between atoms of carbon or hydrogen in the wax and oxygen in the air. As they do so, heat is released which feeds the combustion. This is the corollary of the wife leaving home with her lover. Nipping back and forth between atoms of carbon or hydrogen from the wax and oxygen from the air, the electrons form the *covalent bonds* in carbon dioxide and water. The covalent bond between atoms is like a divorced parent who lives in a new home but visits the old home.

When the silver metal sodium reacts with the green gas chlorine an explosive reaction occurs. Excited electrons leap from the sodium atom into the chlorine and never return. The sodium is left with a positive charge and the chlorine, having acquired a new electron, gains a negative charge. These charged atoms are called *ions.* The electric charge between the ions of sodium and chlorine sets up an *ionic bond* between them. In solution, sodium and chloride ions drift about continually bonding and breaking up.

In the human situation ionic bonding is analogous to people in explosive relationships departing for new homes and never returned to the old ones. Breaking free of a charged situation they find themselves in the situation of drifting from one temporary liaison to another; which is precisely what charged ions in solution do!

The parallels in the vortex theory between quantum interactions and human relationships are remarkable and this degree of Hermetic, *as above so below: as below so above,* symmetry speaks for the validity of the thesis.

There plenty of theories in physics and cosmology and many have appeal. However, for a theory to serve the population at large it has to be simple and understandable and relate to people. People also expect a theory to be believable and a standard measure for the believability of a theory is its ability to explain thingsthat other theories cannot.

I believe a true scientific theory should also satisfy philosophy as wellas physics. The mechanics of the quantum vortex suggests subatomic particles cannot move faster than the speed of light because waves and vortices interacting in the quantum world are particles *of* the speed of light. Obviously particles of movement cannot move faster than the speed that forms them. However that does not mean all energy in the Universe is constrained to the speed of energy that make up the particles we perceive. Energy could exist elsewhere in the Universe, beyond our perception, at speeds beyond the speed of . No experiment in physics proves otherwise.

In his book 'Faster than the Speed of '[1] the physicist Joao Magueijo puts forward good reasons to assume that does exist in the Universe with faster speeds than we perceive. According to Dr Manjir Samanta-Laughton[2], in an interview with Robin Williams in 2003, he spoke of his belief that the speed of is infinite and different speeds are expressed through different dimensions of the Universe. I take this idea further by applying the quantum principle. Applying the Hermetic symmetry of *as below so above*, I suggest that critical speeds of energy define dimensions in the Universe in much the same way that critical frequencies of energy define the quantum states in the atom. These levels of energy speed would correspond to the *planes* postulated by Hermes. It is this *quantum leap* in thinking that has enabled me to reconcile spirituality with science and reintroduce the concept of

1   **Magueijo** J. *Faster than the Speed of* Arrow 2004
2   **Samanta-Laughton** M. *The Genius Groove* Paradigm Revolution Publishing 2009

soul as described in my book, 'Continuous Living'[3] that was dismissed for science by Democritus in his material hypothesis.

3   **ASH** D. *Continuous Living in a Living Universe* Kima
      Books 2015

Chapter 8

Nuclear Energy

Nuclear energy… we have the formulas for that, but we do not have the fundamental laws. We know it's not electrical, not gravitational and not purely chemical, but we do not know what it is.

Richard Feynman

Nuclear energy, derived from the transformation of mass into heat, light and sound is a mystery which begins with the stability of protons. How can protons that are trillions of years older than the Universe, suddenly release explosive energy from an apparent breakdown of their mass?

If mass is vortex energy and the spherical vortex is a very compact form of that energy then if the vortex were to unravel, mass could release vast amounts of energy. But protons are infinitely stable. They don't appear to unravel! Nuclear binding is also a mystery. If things are glued together one would expect the more glue the stronger the bond. With the binding of protons in atomic nuclei it is very different. As the mass that glues them into the atomic nucleus is transformed into light the binding gets tighter. Glue is lost and the bond gets tighter; that makes no sense at all!

The mesons discovered by Cecil Powell could hold the clue to solving these mysteries. Protons could contain mesons which are unstable vortices. They account for so maybe it is the unraveling of mesons that accounts for nuclear energy. Cosmic ray energy forced through the nucleus of an atom caused mesons to appear. That suggests there is space in the nuclear particles to accommodate them. In the 19th century people believed the atom was solid then Lord Rutherford showed it is mainly space. Now mesons reveal that subatomic particles in the atom may be full of space too. It makes sense if they are vortices.

85

Think of a vortex as candy floss! At fairgrounds, bundles of candy floss are formed when a stick is plunged into a spinning tub of hot sugar. Strands of sugar caught on the stick are drawn round in a spiral path to form an open, lattice-like structure.

With internal space any radiant energy traveling into the vortex would find itself caught in a spiral. Swirling in the spiral space inside the proton or neutron vortex, the wave-train energy would be transformed into mass.

Imagine the captured swirl of energy filling the space in the vortex. It would effectively transform the candy floss into a candy ball; an apparently hard, solid sphere with a defined diameter, precisely how protons and neutrons – collectively known as nucleons – appear to be.

The Crab Pot

With an ever tightening spiral path of internal space, nuclear particles would be as energy traps. One could liken a proton vortex to a crab pot!

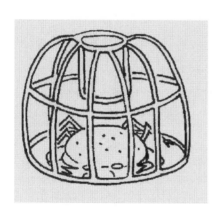

Because of the way the crab pots are made the crab can enter the pot but then it can't escape. In the same way, the spiral structure of space within the proton would make it easy for energy to enter the vortex but difficult for it to leave again. The captured energy could end up swirling around indefinitely inside the nuclear vortex.

Just as a crab pot takes no action to capture crabs, so a proton or neutron vortex would not act to capture energy. The pot is passive. It is the crab that crawls into it. In the same way, the proton vortex would be passive. The radiant energy would move into it and become captured.

A trap can continue to capture creatures until it is full. In the same way, a proton or neutron vortex could continue to capture energy until in the fullness of time it is full of captured energy. I call captured energy spinning in a nuclear vortex *captured mass.* I believe the meson is captured mass and captured mass is the origin of nuclear energy.

Complete Capture

A proton is nearly two thousand times as massive as an electron so it has far greater passive inertia. The inertia of a proton or neutron vortex would enable it to withstand the impact of an incoming packet of radiant energy. This would enable the entire wave train to drive into a nuclear vortex so all of its energy would be *completely captured* and transformed into mass.

Partial capture could occur after the massive vortex is saturated

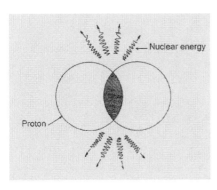

with completely captured energy. Then it could be propelled into wave motion – much like an electron – by energy driving into the more peripheral region of the vortex but the active inertia of the bombarding energy would have to be sufficient to overcome the passive inertia of a proton. In the intense heat of the sun protons have sufficient kinetic energy to attain very high velocities, enough to collide with an immense force, con-

verge. If proton vortices collide and converge their capacity for captured energy would be reduced so some of it would be displaced. Reverting to its radiant form it would fly away. That could explain nuclear energy.

Protons saturated with captured energy can be likened to buckets full of water. If one bucket is placed inside the other its capacity to contain water would be reduced and some of the water would spill out.

Nuclear Fusion and Fission

The capture of energy by a proton vortex causing it to be transformed into mass fits with the first quantum law of motion. The release of mass as nuclear energy, when protons converge, in the sun or thermonuclear explosions, accords with the second quantum law of motion. If the energy caught in complete capture in a proton or neutron constitutes the meson, the loss of some mass in *nuclear fusion* would come from the decay of mesons not protons. As captured energy, the meson mass would have an innate tendency to revert to radiant energy if any of it were to escape.

To explain nuclear fusion, as protons converge their residual captured energy could swirl between them. The swirling of captured mass, around the centers of the two or more nuclear vortices, could fuse them together. This is the vortex account for the strong nuclear force, the force of binding of nucleons in the nucleus.

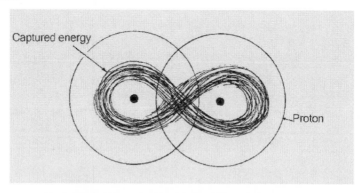

So why does nuclear binding increase when the binding mass drops? That is simple to explain! As nuclear vortices converge closer, the distance between their centers would decrease so the captured energy would swirl in a shorter circuit binding them more tightly together. Any increase in convergence would displace more captured mass, releasing more nuclear energy. At the same time the decrease in the race of remaining energy would further tighten the nuclear particles together. This explains why, after the release of nuclear energy and loss of mass following nuclear fission, binding between nuclear particles is increased.

The Hedgehog Model

To understand nuclear energy and nuclear binding imagine a proton vortex as a hedgehog and captured energy as its fleas. Just as every hedgehog carries a resident population of fleas so every proton would carry a population of captured particles of

energy. The hedgehog never did anything to acquire its fleas; they just hopped on board. So protons do nothing to acquire captured energy; the wave-trains of energy simply drive into them. If a hedgehog is weighed, the greater part of its weight would be that of the hog, but a lesser part would be that of its fleas. In like manner the greater part of the measured mass of a proton would be that of the proton vortex, but a lesser part would be the swirling captured meson energy it contains.

The hedgehog prickles represent the charge repulsion between protons. Because of their prickles hedgehogs would have to be forced to converge. So it is because of their charge repulsion protons only converge if they collide with force. As the prickles of the two hedgehogs are pushed together there would be less space between them for fleas and so some of the fleas would be evicted. In the same way, as two protons converge there would be less space within them for captured energy and so some would be lost and radiate away as nuclear energy. With the loss of fleas, the weight of the converged hedgehogs would be less than the sum of their weights before they were slammed together. So the mass of converged protons would be less than their masses before they collided. Most of the fleas would remain on the hedgehogs and not being bothered about which back they bite, they would hop from hog to hog. So captured energy would circulate between the converged protons and bind them together. In Alice in Wonderland, the game of croquet where hedgehogs were used as croquet balls could be viewed as a premonition of the process of nuclear fusion.

Chapter 9

Electromagnetic Forces

When the solution is simple, God is answering.

Albert Einstein

The vortex of energy provides a simple account for the forces associated with primary particles of matter. Forces are effects of matter that extend in three dimensions and that is the clue that links the forces of electric charge and magnetism to the vortex. Three dimensional extension is a characteristic of the vortex because vortices are three dimensional spirals. The vortex confers three dimensional extension on matter and the 3D extension of the mysterious forces associated with matter – enabling matter to act at a distance – suggests that they are caused by the spherical subatomic vortex.

Action at a distance between bodies of matter has always perplexed mankind. Electric charge and magnetism act over a distance. How can things act on each other when there is nothing but space between them? Early scientists conceived of force fields permeating empty space. Einstein then suggested that gravity was a property of space itself. The vortex theory of energy provides a simple and obvious explanation for force fields.

Electric Charge and magnetism

Electric charge and magnetism have been observed over many thousands of years. Charge causes dry hair to be attracted to a charged comb and bar magnets bounce off each other without touching. As far back as 600 BCE, the Greeks discovered that if amber was rubbed with wool it would become charged and attract lightweight objects. The word electricity is derived from the Greek word 'elektron' for amber. An electric charge can be imparted on many objects by friction. Sometimes when a shirt is pulled off static occurs, caused by the accumulation of electrons that have been stripped by friction off atoms in the material.

The word magnetism came originally from the name given to an ore of iron found near the ancient city of Magnesia. Fragments of this ore were observed to attract small pieces of metallic iron. It was found that this effect was most pronounced at particular areas on the stones, which came to be known as their poles.

As early as 121 CE, the Chinese used lumps of magnetic ore to magnetize rods of iron. When suspended, these rods aligned themselves in the north-south direction and suspended magnets were used as aids to navigation in the West from the 11th century onwards. Because natural magnets were used for direction finding, they came to be known as lode stones from the old English for 'way'.

Electro-magnetism

In 1820, the Danish scientist, Hans Christian Oersted (1777-1851) first demonstrated a link between electricity and magnetism when he showed that an electric current in a wire could deflect a magnetic compass needle. Later, the English scientist, Michael Faraday (1791-1867) showed that a moving magnetic field would induce the flow of an electric current in a wire. This suggested that electricity and magnetism could be incorporated in a single theory. The Scots mathematician, James Clerk Maxwell then developed the equations for electromagnetism and went on to predict that light consisted of electro-magnetic fields radiating through space. This was based on his hunch that space was full of electricity. Next the German scientist, Heinrich Hertz (1857-94) was directed by Helmholtz to test Maxwell's equations for his doctoral thesis. He produced electromagnetic waves which fitted Maxwell's predictions so the electromagnetic theory was accepted.

Maxwell's Electromagnetic theory was one of the few theories from 19th century physics that survived into the 20th century.

Hertz's experiment showed that alternating magnetic and electric fields radiate waves of energy. These radio waves, in their turn induced a flow of electricity in a wire - the principle behind radio and television. That proved energy in the form of light and radio waves can interact with the electric and magnetic fields associated with particles of matter. Maxwell was an outstanding scientist and the model he developed for light, as a pair of waves moving at right angles, has been an immense contribution to physics.

Throughout the 20th century, fueled by the wave-particle theory of Louis de Broglie and Maxwell's electromagnetic theory, physicists including Einstein, endeavoured to unify matter and light in the same field theory. With light acting as though it were formed of particles and matter behaving like waves it is hardly surprising that most scientists followed that direction. Even though Einstein failed to develop a unified field theory physicists today are likely to resist the idea that matter and light are fundamentally different.

A clear distinction between light and matter is not obvious as naturally occurring wave and vortex particles are constantly combined. Field theories that attempt to unify matter and light do not take into account the possibility that particles of matter and particles of light are different forms of energy occurring in a state of perpetual interaction.

The vortex and electromagnetism

In the vortex theory the electromagnetic forces of subatomic particles of matter are thought to be due to vortex interactions. Vortices of energy are dynamic. When they overlap they interact and these vortex interactions are used to explain the forces of electric charge, magnetism and gravity.

The vortex account for forces begins with the assumption that energy is neither created nor destroyed. Therefore there is no limit to the extension of a vortex of energy. Vortex energy can diminish in its intensity but never vanish altogether. This is illustrated by the principle that one can never get rid of a pie by dividing it into smaller pieces; it just gets more dispersed. As the concentric spheres of vortex energy expand, the intensity of energy in them would diminish but would never go to zero. Like the pie, vortex energy spreads out more thinly but never ceases to be.

To appreciate vortex interactions imagine you are in a quantum vortex and experience it as a fireball. As you move outwards from the fiery center, the energy rapidly diminishes. Suddenly it vanishes altogether, as though you have come to the end of the vortex and broken out of its fiery domain. However, the apparent surface of your vortex is not its boundary; rather it is merely the last intensity of energy that you can perceive. Looking out from this point into the darkness, you can see other fiery vortices moving about. You assume that all these spinning balls of light are separate vortices, occupying a void of darkness – like stars in the night sky. However, your senses deceive you. The apparent void of darkness is full of energy extending from all the vortices. As the invisible energy from one quantum vortex overlaps that of another, there is an interaction between them. Standing on your vortex, seeing nothing but a void between you and the others, you are perplexed at the inexplicable attractions and repulsions between them and speak of *action at a distance.*

As vortices of energy, sub-atomic particles of matter have no bounding surface so every vortex is overlapping every other vortex in existence. (The defined diameter of a nuclear particle represents the limit of a proton vortex to completely capture energy; which explains why the strong nuclear force is limited to the nucleus.)

Vortices of energy are intrinsically dynamic and because of their infinite extension they always overlap and so are in a continual state of interaction. The interactions between vortices of energy, extending beyond our perception, account for the ability of particles of matter to act at a distance. It is the infinite extension of subatomic vortices of energy that sets up infinitely extending fields of force.

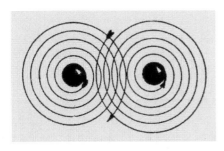

In the vortex theory electric charge is thought to be set up by the flow of vortex energy growing or shrinking in the vortex spin that forms a sub-atomic particle. Opposite direction of spin, in or out of the vortex centre is opposite sign of charge. This would correspond to *Yin* and *Yang* in Chinese philosophy.

Magnetism

Magnetism is thought to be due fundamentally to the rotational spin of the vortex particle. That explains why a magnetic moment is set up by a spinning charge. The in-or-out *whirlpool* type spin forming an electric charge is at right-angles to the round-about *top* like spin setting up magnetism. That explains why a field of electric charge is perpendicular to a field of magnetism. (Electric charge and magnetism are detailed in an appendix on electromagnetism.)

In the vortex theory the force fields of electric charge, and magnetism, caused by interactions of overlapping vortices of energy, contribute to the illusion of material. We perceive matter through interacting forces. In a solid object we experience the rigidity of atoms held together by strong forces between their charged particles. In a liquid the bonds between the atoms or molecules are weaker so it is fluid. In a gas the forces of attraction between molecules are much weaker so they offer least resistance.

As we touch something we perceive the repulsion forces between electric charges in the object and the atoms of our hand. We experience the vortex of energy, not as a material entity, but through its interactions with other vortices. Material substance is explained away by the inertia and extension of the vortex of energy, also by the interplay of wave-trains and vortices of energy forming atoms and molecules and by the vortex interactions setting up the chemical bonds that make things appear to be solid. What seems to be solid and substantial is but the interplay of vortices of energy.

In their behaviour as electric charges, subatomic particles appear like three dimensional ripples. Imagine ripples in a pond going out in all the directions of a sphere. That is the vortex of energy. The concentric spheres of energy expand out of or shrink into a single central point. This model of the vortex of energy helps account for space and time and the infinite extension of the three dimensional vortex of energy also accounts for the discovery in Quantum theory that subatomic particles are everywhere all at once and always connected. It can explain why paired particles can move in concert even when separated by great distances.

Chapter 10

Space and Time

My religion consists of a humble admiration of the illimitable superior spirit who reveals himself in the slight details we are able to perceivewith our frail and feeble mind.

Albert Einstein

Most people assume that space is an emptiness occupied by things but Albert Einstein had an altogether different take on space. Einstein's original thinking about space began when he was five. He was recovering from a cold and his father gave him a compass to play with. As he turned the compass, the needle kept pointing in the same direction. It struck young Albert that space must be holding the needle. Playing with the compass, he conceived of the idea that space was something real, as real as matter itself. It was this that led him later in life to develop his theories of relativity.

Einstein believed matter and space were connected. This is clear from a cryptic remark he made when he arrived in New York, in 1919. Asked by a reporter to explain his theory of relativity in a single sentence, he replied: *"Remove matter from the Universe and you also remove space-time."* [1]

In Einstein's theory of relativity space and time as well as mass are relative to the speed of light. Because mass, space and time are connected and are all relative to the speed of light, space and time appear to be related to the vortex of energy. Like matter, space is a three dimensional extension and that is a characteristic

---

1   **Clerk** R.W. *Einstein: His Life & Times* Hodder & Stoughton 1973

of the vortex. That is another clue why space could be a form of vortex energy.

As energy extends in three dimensions from the centre of a sub-atomic vortex, its intensity would diminish rapidly to infinitesimal levels but would never reach zero because zero is approached but never reached by dividing something into ever-smaller fractions. Imagine you had a balloon that never burst. If you kept blowing it up, the rubber would become infinitesimally thin and the balloon would become infinitely large but the total amount of rubber would always remain the same. It would simply be stretched over a larger and larger area. In the same way, vortex energy would never vanish into nothingness, even though it extends out into infinity. That is because energy is neither created nor destroyed. It follows that every vortex of energy would be as big as the Universe.

Space

Space is an infinite extension and it could be that space is the extension of vortex energy into infinity. That is clear from the vortex theory for force fields. It could be there is no difference between electric charge and space. According to the vortex theory space is electrically neutral only because there are equal numbers of opposite charges in existence that cancel each other out.

As the child Einstein guessed, forces are linked to space. He went on to describe the force of gravity as a feature of space. Space, gravity, electric charge and magnetism are infinite extensions in three dimensions. Maybe this is because they are all infinite extensions of the same vortex energy. It could be matter is vortex energy we perceive and space is vortex energy beyond our direct perception. Perhaps space is thin matter and matter is thick space. Maybe matter and space are the same thing but we have given them different names because they appear to be different.

The blind men and the elephant

There is a story in India of a group of blind men who came upon an elephant. One felt the tail and said it was a vine. Another felt the leg and declared it to be a tree. A third felt the trunk and argued it was a snake whilst the fourth feeling the ear said it was a leaf. None of them had ever seen an elephant so they had no idea what an elephant was. They thought the things they discovered were different and disconnected. They had no idea they were different aspects of the same thing.

Einstein was like man with sight who could see the whole elephant and realised the seemingly different parts were all connected. He realised they were different aspects of the same thing. He discovered the connection between space-time, forces, mass and light and this connectedness was the core idea in his theory of relativity. With the vortex throwing light on Einstein we can appreciate space, forces and matter as the same thing. Mass is our perception of vortex energy through inertia. Forces are our perception of vortex energy through interactions and the void of space is the way we describe the extension of vortex energy beyond perception.

There is no separation in the Universe. Separation occurs only in the mind of man. Realising the connectedness between space, matter and the forces of nature we can come to realise our own connectedness with all that is.

If the vortex is responsible for both matter and space, because matter is divided into particles then space must also be divided into parts. The dense centre of the vortex would constitute the particle of matter and the sparse peripheral regions of the vortex would constitute a part of space. If a vortex were destroyed then both a particle of matter and a particle of space would disappear from existence. This is clear from Einstein's comment: *"Remove matter from the Universe and you also remove space-time."*

If space is vortex energy it must have mass. The mass and structure of space may help us understand dark matter.

If space is vortex energy it must be particulate and particulate space or *space foam* is a prediction of string theory. It is interesting how in the vortex theory, which is a simplistic string theory, the same conclusion is reached.

Bubbles of space

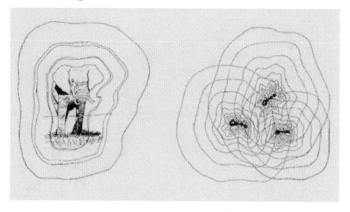

I visualise bodies of matter as being surrounded by *bubbles* of space occurring as an extension of their shape. An elephant, for example, would be surrounded by a bubble of space shaped like an elephant. Its concentric spheres of space would be made up of the particles of space extending from the vortices in the atoms and molecules in its body of matter. Its space would contribute to the foam of Universal space.

According to the logic of materialism we would expect there to be more space in an empty room than in a room full of people. Einstein's relativity seems to suggest the opposite; there would be more space in a room full of people than in an empty room. The vortex theory accords with Einstein as each person would bring into the room a bubble of space extending from his or her body. The space in a room full of people would be denser than the space in an empty room. If you feel someone coming up close to you is invading your space; you are absolutely right!

With distance from a body the intensity of energy contributed by its bubble of space would diminish rapidly. However, the summation of the bubbles arising from neighbouring particles would contribute to the intensity of space in that region. The bubbles of space reaching out from every body of matter in the world would contribute to the concentric spheres of space extending from the Earth. Adding to the space of the Sun and other planets, stars, and galaxies, the space bubbles extending from every subatomic vortex in matter would add up to form universal space.

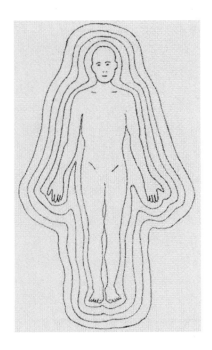

Each one of us is surrounded by our own bubble of space which extends throughout the Universe, overlapping everyone else's space. Every thought we think could pass as a vibration through our individual, infinite *aura of space* influencing every other human through their own extending *space aura*. It could be those who think similar thoughts to you will resonate to your frequency of thinking and be influenced by your thoughts. None of us are islands. We are all connected through space and could have subtle *space effects* on each other. Understanding space to be a real form of energy, as real as matter, allows for many possibilities.

Bell's theorem[2], verified in 2014, fits well with the theory for 'bubble' space or 'space foam' because it is based on the infinite extension and connectedness of vortices of energy. The connectedness of quantum vortex particles explains the principle of *quantum entanglement,* that paired particles once connected then separated by a distance, can still operate instantaneously inconcert,

Space bubbles would share, with elementary particles of matter, all the properties of the quantum vortex. This is why – as perceived by the child Einstein – forces such as gravity, electric charge and magnetism could be viewed as properties of space.

As a form of vortex energy, space could be electrically active. The apparent electrical neutrality of space could be a consequence of the numerical equality of opposite charges in existence where electrical activity is cancelled by the mutual co-existence of opposites; as depicted in the atom.

Magnetism can also be perceived as a property of space. The Earth's magnetic field can be viewed as a property of the space extending from the . Einstein developed his theories of relativity from the idea that a compass needle was held by space. His success suggests this idea is correct.

The occupation of space by matter and fields of force is the ability of one vortex to occupy another. Every bubble of space occupies all others. This is the essence of relativity. Everything is relative to everything else not to some arbitrarily absolute space and time. For example, the and her magnetic field occupy the space extending from the sun, the moon and other planets in our solar system and other stars in our galaxy. Because they are distant from us the magnetic and gravitational effects of their space

2   **Bell** J., On the Einstein, Podolsky Rosen Paradox.,Physics 1(3) 1964

may be less than our own but significant enough to impact the weather, the ocean tides and our own biorhythms.

The Riddle of Relativity

In his special theory of relativity, Einstein predicted that space and time are relative to the velocity of light. But how can space and time be relative to a speed of movement when speed is itself a relationship between space and time? It is a chicken and egg dilemma, what came first, space-time or movement? This 'riddle of relativity' is resolved in the vortex theory by treating one vortex of energy as a particle of matter moving relative to another vortex acting as space. No particle nor man or woman exists in isolation. All bodies and their extending bubbles of space move and exist relative to everything and everyone else. Every being and body in the world live in the vast womb like bubble of space extending from the Earth. Just as we are all held by the gravity of the Earth we also move relative to her space. Every vortex of energy acts as a system of movement relative to every other vortex existing as space. These are all relative to the speed of light because that is the underlying speed of motion in them all. This relativistic principle resolves the dilemma of how a vortex of energy can both extend in three dimensions and move three dimensions. Each vortex extends and moves in the three dimensions set up by all other vortices of energy.

The vortex theory points to unity between quantum theory and relativity theory. Vortices of energy as particles of motion are spin at the speed of light. Each vortex of motion exists relative to all other vortices which suggests that everything in the Universe is interdependent i.e. *the whole is made of parts and every part relates to the whole.*

The law of love

I view the inter-relationship and the interdependence of all things at a quantum level as the universal *law of love*. It makes no sense for us to destroy each other and other things if everything depends on everything else to exist. It makes sense to support and care for other beings if we and they are totally inter-depend-ant. Most of all we should care for the Earth because we are ut-terly dependant on her for everything from space to move in and gravity to hold us along with the air we breathe, the food we eat and the water we drink.

These principles, fundamental to the vortex theory, are the foun-dation of the Yogic tradition. Yoga means *union with all that is.* The way of Yoga is to seek union with the whole of which we are all a part. We can begin to see the wheel of Yoga in the vortex the-ory. Through the vortex theory we can see common threads be-tween modern physics and ancient Yogic philosophy.

Time

Time is the relationship between the parts and the whole. We as parts take our measurement of time from reliable sequences of events in the whole around us. The spin of the Earth sets our days, the motion of the moon gives us months and the orbit of our planet sets the years. Changes in the atom set the most accu-rate time as they are the most dependable events. Every move-ment or change in the Universe exists as a sequence of events relative to the other processes that are occurring around it. The interdependence of events in our world is what we experience as time. Time is one process occurring relative to other repeating physical processes around it. Most significant is the time set up by the passage of light through space. The unidirectional move-ment of light through space sets up the time we experience as the flow from past through the present into the future.

Einstein included time in his theories of relativity. He treated time as a fourth dimension – existing in continuum with the three dimensions of space. Once you appreciate time in terms of things happening relative to you, then you can begin to see time from Einstein's point of view and understand his idea that time dilates when bodies accelerate.

Imagine if you were able to accelerate toward the speed of light. You would begin to notice the surrounding flow of events, which act as your clock, seem to slow down. If others were using the flow of energy in your vortices as their clock then, as a result of your acceleration, to them, their time would appear to speed up.

Two of my daughters are twins. If they were the only beings in existence the extending vortex energy forming one twin would be the space for the other and the flow of vortex energy in one would be time for the other. Each twin would depend upon the other twin's vortex energy for her space and her time. As each of them would be time to the other, if one were to accelerate, from her point of view her twin would have slowed down so time would have slowed. From the standpoint of the stationary sister time would have speeded up. The result is they would begin to age differently. The static twin would age faster than her accelerating sister.

In his theory of special relativity Einstein predicted this extraorinary twin paradox and it has been proved by experiments in particle accelerators.

Muons that have been accelerated last longer than muons at rest[3].

The acceleration toward the centre of the quantum vortex would also effect time. If you were using a vortex of energy as a clock then, with acceleration toward its centre, you would experience a relative dilation of time.

Imagine you were to shrink into the atom. The minutes of your normal experience would first become as hours, then as days and then months. With acceleration toward the centre of the vor-

3   **Calder** Nigel, *Key to the Universe:* BBC Publications 1977

tex seconds of time, measured by people of normal size, would be as years to you.

Now imagine yourself as a space-giant who treats planets as footballs. Leaping from star to star you might look down on the planet Earth spinning like a top and circling around the sun. You could treat the globe as a clock. Human days would count your seconds and years your hours. On a galactic scale, looking down on Earth from the heavens, millions of terrestrial years would appear to be but days.

The Bible story of creation in a week is anathema to scientists who believe in cosmology and evolution occurring over billions of years. Creationists believe the Bible story to be a fact. The understanding in the vortex theory that time is relative to size could help to reconcile scientists and creationists over the time scale of creation versus evolution even if they cannot agree on the mechanisms. Understanding that time is relative to the size of space could help us appreciate that evolutionists and creationists are neither right nor wrong; they are just representing different points of view.

Chapter 11

Gravity

In theoretical physics, the search for logical self-consistency has always been more important in advances than experimental results.

Stephen Hawking

The equations for gravity are identical to those of electric charge which suggests gravity may be a vortex interaction but if so why is it so much weaker than electric charge and why doesn't it appear to have a polar opposite like electric charge? I found clues to answering these questions in *Alice's Adventures in Wonderland.*[1]

Alice went on an extraordinary journey when she went through a looking glass into a world that mirrored her own. She then fell down a rabbit hole following a rabbit obsessed with time. At the bottom Alice had to shrink so she could pass through a tiny door into wonderland.

To me the rabbit hole represents the vortex of energy and the white rabbit represents time. Falling down the hole is the acceleration into the centre of the vortex. Shrinking in size at the bottom of the hole depicts the diminishing

---

1   **Carroll** Lewis, Alice's Adventures in Wonderland, 1865

size of space toward the vortex centre and the tiny door is zero space; the point of singularity at the centre of the vortex. Beyond it lies the wonderland of antimatter.

Antimatter

In the imagery in Lewis Carroll's book I saw a mirror symmetrical world of antimatter existing beyond a point that could be reached only by shrinking in size. The identical twins 'Tweedle Dum and Tweedle Dee' who agreed to do battle represented *mirror symmetrical* twin particles of matter and antimatter with their tendency to annihilate.

The clues to antimatter in Lewis Carroll's story excited many physicists but in the vortex theory they point to a world of anti-matter existing beyond the core of matter. The vortex theory can only work if there is a mirror world of antimatter beyond the centre of our world of matter.

As I developed my vortex theory I had to explain where the energy went to after it reached the centre of the vortex. I envisionedit spinning into the centre as though down a funnel. I imagined it then passing through the centre as through a tunnel and then in my mind I conceived it spinning out again to form another vortex identical in size but opposite in direction of spin.

To account for the origin and fate of vortex energy I predicted that when energy spirals in a vortex to form a negative charged electron it passes through the centre and spirals out again to form the positive charged particle of antimatter called a *positron*. But when energy spirals out of a proton where is the *anti-proton*? To answer that question I went on an imaginary cosmic journey. To begin with the centre of a proton in my head was expanding as a minute bubble. Each sphere of energy grew as another bubble expanded inside it from the proton centre. Concentric bubbles of energy popping out of the point of singularity created a stream of corpuscular spin forming concentric spheres of *spherical vortex energy* that made up the proton.

Journey in Space

In my imagination I joined one of the bubbles and grew with it out into the atom. The energy in my ever expanding bubble became imperceptible beyond the atomic nucleus but the spheres of energy from the other particles in the nucleus merged with it to form a sphere I was on expanding out from the nucleus, just as the original sphere had grown out of the proton. There were larger spheres ahead of me and smaller ones behind, all expanding on the same journey. As my bubble of energy grew out into the atom I could see in my minds eye concentric spheres of energy going in the opposite direction. They were shrinking into the electrons orbiting the nucleus.

I was now surrounded by millions of atoms in a DNA molecule. Then swelling past mitochondria factories, busy membranes and engorged vacuoles, I found myself outside a brain cell. Still growing, the sphere of energy was first as big as my brain and then my body. No longer a sphere it was an extension of my shape and contained vortex energy from every subatomic particle in my body.

My aura of vortex energy merged with the space bubbles of other people as it continued to grow. They shrank then vanished as my bubble of energy grew to enclose the entire Earth. At 13 thousand kilometers diameter the mighty sphere contained vortex energy from every particle in the planet. The Earth was vast beneath me, but then as my sphere of energy continued to expand the globe shrank to a blue and white marble suspended momentarily in space, before she was gone.

My sphere of vortex energy was now an orb, 12 million kilometers in diameter, surrounding the entire solar system. Jupiter and the sun were burning red and gold then vanished as swelling to 100 trillion kilometers, my bubble of space included the three stars of Alpha Centuri, Sirius and Barnard's single star. Still growing, at a million, trillion kilometers the bubble had become a mighty spheroid containing the vortex energy extending from every proton in the entire Milky Way galaxy. Soon the bubble of energy had grown to encompass a cluster of some twenty galaxies including the Milky Way, the Magellanic clouds and Andromeda, bursting with billions more stars.

In my imagination the cluster shrank from sight as my bubble grew to enclose hundreds of galaxy clusters including Virgo with thousands of galaxies and Canes Venatici with thousands more. Expanding out of this super-cluster the orb was 100 billion, trillion, kilometers across. It enveloped the super-clusters, Peruses, Hercules and Indus. At 200 billion, trillion kilometers it encompassed energy from thousands of super-clusters. At 300 billion, trillion kilometers I began to encounter quasars, radiating vast amounts more energy than from stars. At 400 billion, trillion kilometers my sphere had swept up countless millions of quasars and with a diameter of 500 billion, trillion kilometers it was the largest sphere of space containing vortex energy extending from all of matter. For a moment I was on the largest orb of space, at the outermost frontier of the Universe, before it stopped expanding and began to shrink.

Instead of traveling out with accelerating galaxies I was now hurtling toward them. Faint specks of light grew into the mighty super-clusters Peruses, Hercules and Indus that expanded and swept past me. I seemed to be shrinking back the way I had come as Virgo appeared and exploded into a thousand galaxies then as Andromeda flew past I was enveloped once again by the Milky Way.

A single point of light enlarged toward me until it became the full fire of the Sun. Jupiter, Mars and Saturn then swelled past me as I shrank towards the white and blue marble Earth expanding out to greet me. Viewing familiar continents amidst the clouds, I was drawn to England. Dawlish in Devon, with its black swans, appeared as I raced toward a human body. At first it was minute but as my sphere diminished it grew bigger than a giant. Hair shot up like trees as I vanished into a single brain cell then into its nucleus. A coil of DNA appeared and I shrank into the double helix, then into an atom. Spheres of energy were expanding out of its orbiting vortices. They were positrons, not electrons! I had just passed through a world of antimatter! Irresistibly drawn ever closer to the antimatter atomic nucleus I was spinning into an anti-proton until I became a single sphere of energy. Compressed into the point at its centre I slipped through singularity and begin to spin out again. I found myself back in the proton where I had started. I woke up from my dream leaving the white rabbit of vortex energy to chase time through the endless spiral of space between looking glass worlds of matter and antimatter.

The Alician Dimension

My imaginary journey depicted the Universe as two identical and equal halves in which vortex energy spins endlessly between matter and antimatter in the dimension of size of space; which, after Alice, I called the *Alician dimension*.

I liken the Alician dimension to the water cycle. A whirlpool can exist indefinitely in a river so long as water is continually flowing through it. Rivers are fed by water evaporating from the oceans and they in turn feed water back into the sea. An endless flow of water in the river is only possible while rain continually replenishes the cycle. The water cycle exists in two halves. One half is the flow of water from the springs to the ocean through the rivers. The other half is the flow from the ocean to the springs via the rain.

The unending flow of energy in the vortex between matter and antimatter could be part of a universal cycle of vortex energy similar to the water cycle. The innumerable centres of subatomic particles are akin to springs and the one largest sphere of space could be likened to the oceans. The flow of energy through vortices of matter would be analogous to streams and rivers carrying water from where it rises in the springs to its destination in the ocean. The evaporation of water from the seas to form clouds and then rain to maintain the springs would represent the other side of the vortex cycle of energy, the flow of energy from the largest sphere of space back through the vortices of antimatter.

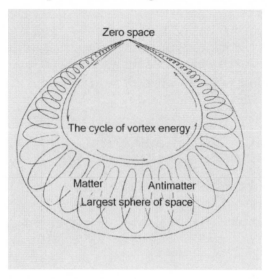

Matter and antimatter

Because we are formed out of the vortices of energy in matter we can only be aware of the matter half of the cycle. Like fish in a river we would know only the watery domain of the river and have no knowledge of clouds and rain above it. Scientists observing particles of antimatter penetrating the world of matter could be likened smart fish observing drops of rain penetrating the surface of the river and predicting another half of the water cycle beyond their limited experience in the river. The discovery of antimatter enabled me to predict two halves of the universe, matter and antimatter, to account for the endless cycle of energy.

Matter is attracted to antimatter so if the cosmology of the vortex is correct, particles of matter in our world should experience a pull through the regions of minimum-space, towards particles of antimatter in the mirror image world. Acting through the centres of subatomic vortices this pull would be a centralizing force causing particles to conglomerate. Coming from another world that is remote from us, the vortex attraction would seem to be weak compared with other forces more immediate to us in our world. But the strong centralizing force on your body could be a pull on you from the enormous mass of antimatter beyond the Earth. Acting through the centre of every particle in the planet this vortex interaction between the worlds of matter and anti-matter could account for the pull on us all from the Earth; the force of attraction we call gravity.

The polar opposite to gravity

There is a polar opposite to gravity. It is a pull on matter from theworld of antimatter, through the largest sphere of space. Acting on the largest bodies of matter this decentralizing pull from the outermost reaches of the Universe of matter would affect galaxies causing them to fly apart from one another. That would result in an apparent expansion of the Universe. As a vortex interaction, akin to electric charge, this pull on the galaxies would be a force of acceleration operating from the furthermost reaches of space. Galaxies that are further from us would therefore appear to be accelerating faster than galaxies that are closer to us. This would account for the accelerating expansion of the Universe.

I first published this prediction as the *cosmology of the vortex* in 1995 [2] which explained why the furthermost galaxies would be accelerating faster away from us than nearer ones. Perlmutter published confirmation of this in 1997 [3]. That is why I can call my

2   **Ash** David, *The New Science of the Spirit,* College of
     Psychic Studies, 1995
3   **Perlmutter** Saul., et al. Discovery of a Supernova
     Explosion at Half the Age of the Universe and its
     Cosmological Implications Lawrence Berkeley National
     Laboratory, Dec 16, 1997

vortex theory a scientific theory rather than an hypothesis. When I started work on the vortex cosmology in 1969 I lacked evidence of an antimatter half of the Universe. Now there is evidence in abundance!

Gamma rays are known to come from the annihilation of matter and antimatter[4]. Recently discovered gamma rays bursts are distant events from outside our galaxy. They are most commonly associated with the core collapse of very massive stars into black holes so gamma ray bursts provide evidence that antimatter exists beyond the cores of stars. As predicted in the vortex cosmology gamma rays come from where mass is most centralized; where gravity is strongest.

Recent discoveries of high speed streams of electrons from the centres of galaxies also suggest that annihilation of matter and antimatter is going on inside the black holes located within them. William Tiller of Stanford University and William Bonner of London University relate this to gravity and the hidden properties of a mirror world [5].

The evidence of antimatter beyond the core of black holes provided by bursts of gamma rays, high energy electrons and X-rays now recorded regularly as astronomical events, add strength to the original predictions in the cosmology of the vortex theory that antimatter exists beyond the centre of matter. Increasingly this model is supported in the scientific journals. In the Scientific American article *The Brightest Explosions in the Universe*[6] an account is given for annihilation between matter and antimatter through the centres of black holes .

4   **Richards**, et al, *Modern University Physics*
     Addison-Wesley 1973
5   **Tiller** W. Science and Human Transformation Pavior, 1997
6   **Gehrels N**. et al, *The Brightest Explosions in the Universe*,
     Scientific American Dec. 2002

The big bang theory failed to predict these events or the accelerating expansion of the Universe and it incorporates a fable masquerading as fact that: *in the beginning there was matter and antimatter but slightly more of the former than the latter. All the antimatter annihilated with most of the matter and the remnant that we see is the Universe.* This speculation endorsed by mainstream science and presented in books, articles and on television, shows inadequacies in the big bang theory and another weakness in the standard theory.

In my 1995 book I suggested the Universe would appear to be expanding from the pull between matter and antimatter rather than a push from a big bang. As a pull is the antithesis of a push I proposed the vortex cosmology as an antithesis to the big bang theory.[7]

Conjectures of an antimatter half of the Universe abounded in the 20th century from Dirac to the 'Trouser Leg Universe' of Tel Aviv University, but were dismissed for lack of evidence of gamma rays. The evidence is now in but some still resist to the idea of an unseen underworld of antimatter because as Thomas Kuhn stressed, *truth has as much to do with the consensus of scientists as to the outcome of experiments*[8].

I find it all deeply frustrating. Rather than allowing time for the advance of science to bring in evidence to support the obvious theory of an antimatter half to the universe, the theory was rejected because of the lack of evidence of gamma rays in space to support it. When the evidence of gamma rays from galactic cores eventually appeared, instead of resurrecting the anti-matter theory, other theories have been supported in the press and in

7   **Ash** David, *The New Science of the Spirit,* College of
     Psychic Studies, 1995
8   **Kuhn** T. The Structure of Scientific Revolutions ,
     University of Chicago Press, 1962.

programmes such as *Black Holes* on BBC *Horizon* to account for the gamma rays and high speed streams of electrons coming from Black Holes, when these energies rightfully belong to matter-antimatter.

Quasars

According to the vortex theory, if a galaxy of matter were to accelerate toward a galaxy of antimatter and vice versa, eventually they would meet and annihilate. If this cosmology is correct there should be evidence of annihilation going on at the largest regions of space where matter and antimatter meet. In the outermost reaches of the Universe the furthermost things from us are 'quasars'. Quasars are quasi stellar phenomena associated with black holes. They radiate in the order of two hundred times the energy of normal stars [9]. Stars produce energy through nuclear fusion, which consumes 0.7% of their proton mass. A process which produces in the order of two hundred times this amount of energy could be consuming all of the mass of a proton. The only process we know of that can do that is the annihilation of matter and antimatter.

Observations that galaxies are accelerating away from us have led to the assumption that the Universe is expanding. In the original publication of my vortex theory I suggested the acceleration of galaxies of matter toward galaxies of antimatter could be a movement toward annihilation rather than indefinite expansion[2]. Are the quasars in the outermost reaches of the Universe galaxies of matter annihilating with galaxies of antimatter?

Quasars exist at either end of the spectrum of space; occurring in the largest as well as the smallest realms of space. These are the regions of densest gravitational contraction and greatest universal expansion where, according to the vortex theory, matter

9   **Greenstein J & Schmidt M** *The Quasi-Stellar Radio Sources.*
Astrophysical Journal 140, 1964

meets antimatter. In the vortex theory quasars are also formed by gravitational collapse that occurs after stars exhaust their nuclear fuel and cool down.

Dying stars eventually collapse in on themselves due to gravity forming a neutron or dwarf star but in the demise of very massive stars, gravity could be sufficiently strong for a black hole to form. Black holes in the centres of galaxies can engulf millions of exhausted stars in their gravitational maelstrom. Our modest galaxy has a moderate black hole at its centre housing a mere seventeen million exhausted stars in a sphere that would occupy the space of our sun out to the orbit of Mercury.

Many galactic cores are much more massive engulfing hundreds of millions of dead stars. But is the black hole an elephant bone yard of stars? Is it monstrous black tomb of death beyond our darkest imaginings or is it the birth of something bursting with light, brilliant and glorious beyond our wildest dreams? The cosmology of the vortex suggests the latter may be true.

Black holes represent the greatest density of matter and intensity of gravity. As a black hole forms in matter, an antimatter black hole could be forming beyond it through the smallest realms of space at its colossus centre. As particles of matter in the black hole are centralized into singularity, they would be converging on antimatter in the mirror black hole through its zero space point and there annihilation between matter and antimatter would occur.

To begin with, the energy released by annihilation would not radiate from the black hole because acting like a gargantuan proton vortex it would completely capture the energy in its spiral space-path. However, with annihilation, the mass of the black hole would be shrinking and its gravity would diminish and so its ability to capture energy would be decreasing. At the same time, due to progressive annihilation, the energy contained by the black hole would be increasing. Eventually there would be a crossover between diminishing gravity and increasing release of energy from annihilation. This would be a threshold where energy would begin to escape and radiate away. Since the 1970s Stephen Hawking proposed that black holes leak energy which comes from matter-antimatter annihilation.

As annihilation destroys gravity a black hole would transform into a quasar. If this happens then a black hole would be a temporary astronomical formation, occurring as a step in the transition of stars into quasars. If every black hole is undergoing transition into a quasar, then according to the vortex cosmology, most black holes should have quasars associated with them. In fact new micro-quasars are being discovered with black holes at their core that are emitting super fast streams of electrons and gamma ray bursts perpendicular to the black hole centre. The bursts appear suddenly and then die away. The energy and behaviour of these emissions is perplexing.

I contend that the periodic bursts result from a tussle between the gravity causing particles to conglomerate and the energy of annihilation tending to blow them apart. To appreciate gamma ray bursts imagine a black hole as an automatic toilet cistern. The energy of annihilation would build up inside a black hole or neutron star until the gravity could no longer contain it and then it would be released in a burst. Afterwards the gamma rays produced by annihilation would be trapped by gravity again but then begin to build up, ready to escape in another burst. Because this periodicity of gravitational energy emissions is also exhibited by geysers I call it the *geyser model for gravitational energy emission*.

High energy electrons streams and gamma ray bursts, confirming matter-antimatter annihilation through the intense gravitational core of black holes, suggests a black hole may not be an astral grave. It may be more a stellar *pupa* in the metamorphosis of *caterpillar* stars into quasar *butterflies*.

In the vortex theory all gravitational energy is treated as a product of annihilation between matter and antimatter. That fits with Einstein's belief that gravitational energy is derived from the destruction of mass. When a body falls due to gravity it releases energy on impact. This energy is derived from the destruction of mass because a moving body has a greater mass than a body at rest. As a body falls in the world of matter, a mirror-image body of antimatter would also be falling, so both would make their impact and release energy simultaneously. There would be an equal minute annihilation of mass in the matter and antimatter as both bodies come to rest and at rest they would be closer to their ultimate, complete annihilation. In order to lift the bodies again, work has to be done by someone in matter and simultaneously by his or her mirror image in antimatter. That would require an equal input of energy to that released when they came to rest.

The fall of every bit of matter on Earth is another step in the inexorable progress toward annihilation. As you pick up an object that has fallen, you use up energy from the sun that was acting to keep the particles in the sun from falling together due to gravity. Eventually, when the sun's reserves of fuel are exhausted, there will be nothing left to keep you alive let alone lift things! The inertia of gravity will take over and all the matter within our local star will collapse toward antimatter and annihilation. Even though the sun is too small to form a black hole and will eventually settle down as a dwarf star, ultimately this would be swallowed by black holes as the whole galaxy consumes its reserves of nuclear fuel and collapses under the awesome influence of gravity.

Annihilation is the ultimate transformation of matter into light. However, with the annihilation of mass comes the demise of space because space and mass are one and the same thing. The energy of annihilation would become trapped in the collapsing space which would force it into an increasingly tight spiral space path. This is a powder keg set for explosion. As the annihilating galaxies collapse toward singularity their residual mass and energy would be compressed into ever smaller space from which the *big bang* explosion into a new cycle of the Universe could occur.

If a big bang did occur in the past then a big bang will occur again in the future to herald the rebirth of space. But big bangs are not the beginning of the Universe. They are more like birth after death or dawn after dusk. In India they speak of the comings and goings of the Universe as the in-breath and out-breath of Brahma. Whatever the process, whether there is a universal big bang or lots of mini-big bangs, there can be no beginning or end to the Universe. With protons existing for trillions of years longer than the Universe it is clear that we are observing but part of a greater process that represents transformation and change on a scale beyond our comprehension. There is no need to presume a creation of the Universe because the energy that forms the Universe was never created nor will it ever be destroyed. Energy exists. All we observe are its unending transformations.

Chapter 12

A Complete Theory

If we do discover a complete theory, it should in time be understandable in broad principle by everyone, not just a few scientists. Then we should all, philosophers, scientists, and just ordinary people be able to take part in the discussion of the question of why it is that we and the universe exist. If we find the answer to that, it would be the ultimate triumph of human reason - for then we would know the mind of God.

Stephen Hawking

Ockham's razor

Stephen Hawking gave a guide to what we should look for in a scientific theory: *"A theory is a good theory if it satisfies two requirements: It must accurately describe a large class of observations on the basis of a model that contains only a few arbitrary elements, and it must make definite predictions about the results of future observations."* [1]

Hawking's first criterion for testing a good theory was originally described by a philosopher, William of Ockham in the 14th century as: *'non sunt multiplicanda entia praeter neccessitatem'*, i.e. "Entities are not to be multiplied beyond necessity." Ockham argued that in a debate on theories we should accept as true the theory that has the least arbitrary assumptions. He used this law of economy of ideas with such sharpness that it came to be known as Ockham 's razor.

Ockham's razor has been employed by numerous scientists, the most famous being Galileo who invoked it in defending the Copernican hypothesis for the heavens.

---

1   **Hawking** Stephen, *A Brief History of Time* Bantam Press 1988

Lord Rutherford, the father of nuclear physics, said the fundamentals in physics have to be simple and the vortex theory offers a very simple model. From the Yogic idea that the smallest particles of matter are vortices of energy the vortex theory has accounted for:

- Mass as quantity of vortex energy
- Inertia originating in the spin of vortex energy
- Potential energy as vortex energy
- 3D extension of matter, forces and space as the 3D extension of vortex energy
- Force fields as interacting vortices of energy
- Infinite extension of forces fields and space as the infinite extension of the vortex of energy
- Electric charge as expanding or contracting concentric spheres of vortex energy
- Magnetism as rotating vortex energy
- Kinetics in terms of interactions between waves packets of energy and vortices of energy
- Wave particle duality as a consequence of the bound state of vortex and wave forms of energy
- Space as the infinite extension of the vortex of energy
- Space curvature as spherical vortex energy
- Time as the relationship between particles of energy
- Mesons as energy captured by nuclear vortices
- Nuclear binding as energy swirling between vortex particles in the nucleus of an atom
- Nuclear energy as the release of captured energy when nuclear vortex particles converge

- High energy particles as unstable vortices formed by the passage of energy through subatomic vortices

- Strangeness as longevity conferred on an unstable vortex of energy by a stable vortex at its core

- Particles of matter and antimatter as subatomic vortices with equal mass but opposite direction of spin.

- Gravity as vortex interactions between matter and antimatter in smallest space

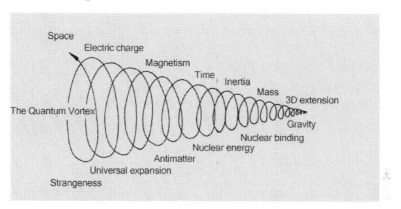

- The accelerating expansion of the Universe as vortex interactions between matter and antimatter in largest space

Predicting the outcome

The second requirement of a good scientific theory is to predict the outcome of future experiments, observations or discoveries. Prior to the 1990's it was assumed that the Universe was expanding at a uniform rate. However, in the 1990's Saul Perlmutter used redundant telescopes to look for supernova explosions in distant galaxies. In 1997 the results from observations of a couple of dozen supernovas were published which suggested the expansion of the is speeding up.[2] Since then more supernova discoveries have supported this conclusion. Decades before Perlmutter's discovery it was clear from the cosmology of the vortex that the further galaxies are from us the faster they accelerate away from us and this was published in 1995[3] two years before Perlmutter published his findings. Predicting the outcome of Perlmutter's observations not only established the vortex hypothesis as a sound scientific theory but made it the only version of string theory that has had a prediction verified by experiment.

The vortex theory has explained classical and modern physics in a way that is simple and self evident so that ordinary people as well as scientists and philosophers can understand it. That alone establishes its value as a theory.

In 1968 my father sent an early draft of my theory to Sir Martin Ryle, whom he knew from childhood. Ryle replied that quantum mechanics provided a satisfactory account for physics and I reminded him of a medieval philosopher working out how many angles stood on a pin head, but I was not deflected. He had

2   **Perlmutter** Saul., et al. *Discovery of a Supernova Explosions...* Lawrence Berkeley National Laboratory, Dec 16, 1997

3   **Ash** David, *The New Science of the Spirit,* Col. Psyc. Science, 1995

missed the point of my work. The value of the vortex theory was that it enabled me as a student to access the quantum world whereas quantum mechanics was so impossibly difficult even professional physicists struggled with it.

Physics is not math

Some people will be concerned the vortex theory is presented without mathematics. The vortex theory explains 'what things are' rather than 'how they work'. Mathematics is better at helping us understand how things work rather than what they are. Math provides formulas but not fundamental understandings. Math may be good for developing a thesis but it may not be so good for inventing a thesis in the first instance. That usually requires vision and imagination and visionaries are not always adept mathematicians and physics is not math. In his book 'The Dancing Wu Li Masters', Gary Zukav said, *"The fact is that physics is not mathematics...Stripped of mathematics, physics becomes pure enchantment."*[4]

Even for physicists, presenting the vortex theory in plain language before math to establish understanding can be an advantage because as Werner Heisenberg wrote, *"Even for the physicist the description in plain language will be a criterion of the degree of understanding that has been reached."*[5]

Quantum tunneling

As for new understandings, in the quantum world there is a phenomenon called *quantum tunneling*. Rather than overcome an energy barrier, subatomic particles can tunnel right through it; a bit like a ghost walking through a wall.

4   **Zukav** Gary, *The Dancing Wu Li Masters*, Rider 1979
5   **Heisenberg** Werner *Physics and Philosophy* Harper and Row 1958

The vortex theory can explain quantum tunneling from the premise that a particle of energy is a bit of speed with a shape. The speed is the relativity constant; represented by 'c'. The shape is vortex or wave. If the speed of energy in a particle were to increase beyond the speed of light without a change in its wave or vortex shape, it could move in and out of space and time as well as move through it.

The Minkowski diagram associated with the theory of relativity makes it clear that speeds faster that of light cannot exist in space and time relative to the speed of light. But if space and time were a consequence of the vortex, energy in the vortex could exist with different speeds and set up different space-time continua to the one we live in. There could be levels of reality beyond the speed of light formed by energy with different relativistic constants.

Our physical world, governed by the speed of light, may be but one level or plane of energy in a series of ascending quantum realities. If the intrinsic speed of energy in a vortex of energy changed, the vortex could vanish out of space and time on one side of an energy barrier and then pop up on the other side. Maybe that happens in the quantum world. Maybe that is how particles engage in quantum tunneling.

The vortex account for quantum tunneling suggests an exciting possibility. If particles can move *in and out* of space and time by change in the intrinsic speed of energy in the vortex why not a body made up of the vortices? If a body could undergo an increase in the speed of energy in every vortex within it, then it could ascend from physical space and time and appear in another space-time reality based on a higher speed of energy, and then descend back into physical space-time at another point.

Inter space travel
Moving in and out of space-time rather than through it allows for the possibility of space travel unlimited by the constraints of distance, time or gravity. Beings in other worlds, somewhere in

the vast, might have mastered the technology of *inter space travel*. They could visit us even if we are not sufficiently advanced to visit them. If there is any truth in the hypothesis of ascending and descending between dimensions of reality through change in the intrinsic speed of energy – the constant of relativity – in the quantum vortex, there could be evidence of visitors in our world from other worlds; beings with the ability to appear and disappear like a tunneling quantum. Maybe this could account for UFOs, apports and a host of other paranormal phenomena.

In his most recent book Steven Hawking wrote:

In the history of science we have discovered a sequence of better and better theories or models, from Plato to the classical theory of Newton to modern quantum theories. It is natural to ask: Will this sequence eventually reach an end point, an ultimate theory of the , that will include all forces and predict every observation we make, or will we continue forever finding better theories, but never one that cannot be improved upon? [6]

A theory that cannot be improved upon would close minds to new possibilities. A complete theory should open minds by allowing for an infinity of possibilities and the possibility of infinity. The value of a complete theory should be not so much in its ability to explain everything as in its ability to predict things yet to be explained!

6   **1. Hawking S & Mlodinow L,** *The Grand Design,*
Bantam 2011

Chapter 13

Consciousness

I regard consciousness as fundamental. I regard matter as a derivative of consciousness. We cannot get behind consciousness. Everything that we talk about, everything that we regard as existing, postulates consciousness.

Max Planck

If there is movement logic demands there must be something moving. This logic led to the material hypothesis of the atom and the ether. By explaining away material substance the vortex theory gives rise to a logical dilemma. How can there be movement if no thing exists that moves? If energy is action what is acting? The solution to this conundrum is simple. Action without an actor is abstraction. Abstraction is in the realm of thought so energy existing without material suggests that energy is more a property of mind than material. If energy is pure movement without underlying substance, particles of energy could be more like thoughts than things. Thought and memory depends on conscious awareness therefore particles of energy could depend on consciousness. This conclusion has enormous implications.

The material hypothesis, embraced by most scientists, holds consciousness to be a product of the brain, particularly the human brain. Recently scientists have conceded that we might share consciousness with other creatures. However, within the ranks of quantum physicists there is a growing number in support of the idea that consciousness may be more universal than we have hitherto been led to believe in science.

Many physicists now consider consciousness to be the bedrock of reality. Amit Goswami, in *The Self-Aware*, concluded that consciousness is the ground of all being.[1] Others share this view. Some write about a universal consciousness while others write books about intelligence as a universal principle. Sir Fred Hoyle, one time president of the Royal Society and one of Britain's greatest cosmologists wrote a book called *The Intelligent*.[2] Other scientists including Fred Wolf, Christian de Quincey, Dale Pond, Peter Russell and William Tiller have written about consciousness as the core principle in the Universe. [3]

Near death experiences

Call it consciousness, self-awareness, intelligence, mind; increasingly scientists are treating cognition as a universal principle.More and more people are coming to believe mind to be a universal phenomenon expressing through chemical processes rather than being a consequence of them. This view is reinforced by scientists writing of near death experience (NDE) such as Penny Sartori[4] and Eben Alexander[5].

NDEs have become an increasingly common phenomenon due to resuscitation. Since best sellers on NDE by authors like Raymond Moody[6], the phenomenon has been widely contested. Efforts have been made to explain away NDE in terms of hypoxia and drugs administered during resuscitation[7] and arguments about the cause of NDE fill the pages of *The Journal of Near-Death*

1   **Goswami** A. *The Self-Aware Universe,* Putnam, 1995
2   **Hoyle** F. *The Intelligent Universe*, Michael Joseph, 1983
3   **Samanta-Laughton** M. *Punk Science* O Books 2006
4   **Sartori** P. The Wisdom of Near-Death Experience, Watkins, 2014
5   **Alexander** E. *The Map of Heaven,* Simon & Schuster, 2014
6   **Moody** R. *Life after Life,* Bantam, 1975
7   **Blackmore** S. *Dying to Live,* Grafton, 1993

Studies. Skeptics argue that NDEs do not constitute valid evidence in the body of science. However, that view betrays a lack of understanding of science. NDEs cannot be excluded from science just because they conflict with the consensus view of what is scientific. Science does not support fashion in belief or points of view[8]. If evidence is elucidated scientifically it is scientific!

Penny Sartori undertook the UK's first long-term prospective study of NDE and published the results of her five years clinical study of NDE as a scientific PhD treatise[9]. The evidence she presented contests the consensus view that the mind is constrained to the brain. Nonetheless it can be treated as valid in science, despite its conflict with current scientific opinion, because she gathered it in a scientific way.

Energy underlies everything. That fact is established in modern physics. The idea that energy rather than material substance forms matter is beyond question in science today. Materialism is a classical theory dismissed by Einstein but whilst most people accept he is the greatest genius in the history of science they do not appreciate his world view that *matter and the field* [light] *are real but they have no substance*[10].

Energy cannot be an act of a material substance if there is no material substance but could depend on consciousness because consciousness exists. We know that for a fact. We may not be able to prove energy depends on consciousness and NDE doesn't constitute absolute proof that consciousness goes beyond the brain,

8   **Popper** K. *The Logic of Scientific Discovery*, Hutchinson, 1968

9   **Sartori** P. The Near Death Experiences of Hospitalized Intensive Care Patients: A Five Year Clinical Study, Edwin Meller, 2008

10   **Berkson** W. Fields of Force: World Views from Faraday to Einstein, Rutledge & Kegan Paul, 1974

but the universality of consciousness can be surmised if particles of energy appear to be more thoughts than things.

Is the universe mind-based?

Reasoning that the Universe is mind based rather than material is sound and in the absence of material some might argue, 'what else could it be?' The quantum laws of motion infer particles of energy are memories. Memories come from thoughts. Thoughts exist in a mind and mind depends on consciousness. There is ample logic to support the belief in the Universe as mind. The postulate that the Universe is a mind with consciousness, thought and memory as prime realities would be supported by physicists and philosophers who treat the Universe more as a mind than a machine and those who treat consciousness as the ground of all being. It is only the groundswell of diehard skeptic materialism, disguised as science, that continues to undermine the conclusion of quantum physicists that there is a universal continuum of Consciousness.

If forms of energy at the speed of light correspond to memory is thought energy beyond the speed of light? Is quantum tunneling interplay between thought and memory in a universal mind? Mind invokes connsciousness. Does consciousness underlie life? One question leads to another, *is consciousness the origin rather than the culmination of evolution?* Consciousness must underlie evolution if consciousness underlies everything.

If the Universe is a mind it would be illogical to exclude intelligence from evolution! Evolutionists contest creationists but the debate, fundamentally, is between mind and material. Because the nature of mind is creative intelligence, creation could not be denied or treated as unsound if the Universe is treated as a mind. These ideas are exciting and speculative but in time the study of creative consciousness and intelligence in evolution may become part and parcel of biological science. Certainly intelligence in every form of life is becoming increasingly self evident!

Observation

Mystics in ancient India proposed a single source of mind giving rise to energy which coalesces into matter.[11] The source would correspond to consciousness if the Copenhagen interpretation of quantum reality is correct; that things exist only to the extent they are observed,. The Universe exists so it must be observed. Observation presupposes consciousness and consciousness is the observer of thought. Yogic thought appears to fit quantum thinking like a glove.

If the distinction between energy and consciousness is between action and awareness of action; between the observed and the observer, and if there is no substance underlying either the observer or the observed then neither one could exist without the other. With no underlying pre-existing substance the observer could exist only to the extent that it is observing. Without the observed it could not be. By the same token, with no underlying substance the observed could exist only to the extent that it is being observed. Without the observer neither could it be. Without underlying substance, observer and observed would be utterly codependent.

The dream state

To appreciate this point, imagine consciousness as a dreamer and energy as the dream. No *thing* exists that is dreaming and no *thing* is being dreamt. All that exists are the states of dreamer and dream. The dreamer exists only to the extent it is dreaming and the dream exists only by being dreamt. Neither state can exist without the other. They are distinct and yet inseparable. The Universe could be the inseparable relationship between consciousness and energy; imaginer and imagination; audience and actor.

11   **Ramacharaka** Yogi, An Advanced Course in Yogi Philosophy, 1904

If particles of energy exist as thought and memory in the awareness of a universal consciousness they may be many but consciousness would be one. The unity of consciousness is clear from an understanding of protons. All protons have identical characteristics such as mass and charge. If these particles of energy are more thoughts than things their identical properties suggest they are thoughts in one mind with a single observing consciousness. If the universal mind and consciousness were particulate then each energy expression of the mind and consciousness would be unique with different characteristics. Uniqueness comes from diverse combinations and interactions of waves and vortices of energy and the unlimited range of frequencies available to wave forms (quanta) of energy.

The great debate for humanity is between materialism and consciousness. Energy is divided but the conscious awareness underlying energy is one. That idea is of immense importance in the debate. The opposite nature of energy and consciousness sets up the distinction between the states of observer and observed. There are innumerable quanta of energy, countless atoms, unimaginable grains of sand in a desert and stars in the heavens but a single conscious awareness underlies them all.

Thoughts are many but the mind is one. Particles are many but the Universe is one. If the consciousness in each of us, seeing through our eyes, hearing through our ears, and feeling through our senses is one then it would be the same one consciousness in all the different bodies of humanity. We may be living separate lives but the thing we have in common is being human. The unity of consciousness we share suggests we are *one being in many bodies*. We may think different thoughts perform different actions and have different experiences but the conscious awareness underlying every thought we think and action we perform is one and the same; that indeed is the brotherhood and sisterhood of humanity.

Some will ask, 'Is universal consciousness God'? In religion God is defined as spirit. Whatever God is, God could not be Source Consciousness if God is spirit as spirit appears to be a very real form of energy and it could be filling space!

Chapter 14

Life

Today there is a wide measure of agreement that the stream of knowledge is heading toward a non-mechanical reality; the universe begins to look more like a great thought than a great machine. Mind no longer appears as an accidental intruder into the realm of matter; we are beginning to suspect that we ought rather to hail it as the creator and governor of the realm of matter.[1]

James Jeans

We live in a world of atomic matter. Vortices of energy in atoms, occurring in three states of matter – solids, liquids and gases – dominate our experience of reality. Most of us are familiar with the picture of the atom and its electrons in spherical or elliptical orbits surrounding a central nucleus of protons and neutrons.

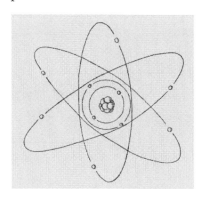

However, vortices could occur in arrangements other than atoms and non-atomic matter is more common in the Universe. Non-atomic matter and ionised atoms form the *plasma* in stars where temperatures are too high for atoms to form. 98% of the mass of the solar system is plasma in the sun so most of the matter in the system is non-atomic. This could apply to the Universe as a whole. Non atomic matter in stars is hot plasma but some plasma is cold and

1   **Burr**, H.S. *Blueprint for Immortality*, Neville Spearman, 1972

135

very little is known about it. Cold plasma consists of electric charged particles blown into space. Through stellar wind and flares, stars blow off electrons, protons, neutrons and atomic nuclei into space with a wide range of energies and space could be accumulating cold plasma much as the sea accumulates salt.

Plasma is cold in space and hot in stars and hot stellar plasma emits light whereas cold space plasma neither emits nor reflects light so it is *invisible*. Cold plasma could account for a considerable amount of mass in the Universe and invisible cold plasma in space could be a linked to life.

No one knows for sure what life is. Some believe it is purely mechanical; a consequence of haphazard interactions between chemical substances based on carbon. Others believe that there is creative intelligence behind life. Everyone has their own opinion but if particles of energy are more like thoughts than things and if consciousness is the bedrock of reality then it is not unreasonable to suppose that intelligence underlies life because life brings order out of chaos. It opposes entropy. It leads to the organization of atoms into complex molecules, cells and multicellular organisms.

In biology we learn what life does not what life is because science is more concerned with what things do than with what things are. Biology has definitions for life such as growth, locomotion, respiration, reproduction and so forth, but these describe functions of life not life itself. It is through physics we can begin to understand the nature of life itself.

A clue to understanding life is water. Biological life depends on water but why does life depend on water? What is it about water that enables it to support life? The answer, I believe, is electricity!

Electricity depends on charged particles. Most atoms are electrically neutral because they contain an equal number of oppositely charged particles but they can become electrically active by gaining or losing electrons. When that occurs they become *ions*.

Hydrogen bonding

Some molecules are charged. The water molecule is charged and electrically active because the electrons involved in the molecular bonding that maintains it as a liquid, are more strongly associated with one atom than another. Water is formed of an atom of oxygen combined with two of hydrogen. The combination is formed by electrons from the hydrogen atoms orbiting the oxygen which 'hogs them' so to speak. The result is that water molecules have an *electric dipole*, a predominance of positive charge over the hydrogen atoms and negative charge over the oxygen. The hydrogen atoms then act as a pair of electric charged prongs that enable the water molecule to undergo *hydrogen bonding*.

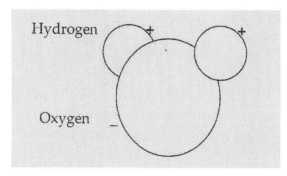

Hydrogen bonding enables the water molecule to bond with other water molecules. It also enables water to dissolves ionic solids like salt and hold oppositely charged ions like sodium chloride in solution which further increases its electrical activity.

Biological life depends on the electrical activity of water. This is evident in the cell because cells contain proteins that form a gel which can hold water in suspension. The water held in the cell enables electrical activity. That makes it possible for life to operate within it. Life on Earth depends on water because water, liquid at temperatures on the planet, is a medium in atomic matter that supports electrical activity.

Plasma

Plasma is electrically activity; in fact it is *the* most electrically active form of matter. Plasma is also the most abundant form of matter in the Universe. If plasma is more abundant and more electrically active than atomic matter and life depends on electrical activity, then life could exist in the invisible plasma of space and be more abundant in space than in the atomic matter of the Earth. The Universe could be teeming with invisible life. We are searching for biological life on planets that are either too dry or too hot or too cold to support it when what appears to us to be empty space could be full of non-biological, *plasmic life* we cannot see.

In the human brain, mind operates through electrical activity. The Universe could be acting like a brain with mind operating through the electrical activity of cold plasma in space. Space plasma offers a mechanism for mind to operate throughout the Universe. The operation of mind in our brains could be a hologram of a universal principle. If space plasma offers a means for mind to express through electrical activity, is that reflected in biology? Is the entire Universe alive and evolving? Is the evolution of life on Earth a hologram of this?

The human brain could be a hologram of a universal brain in which galaxies are as cells and plasma in space is acting like dendrites interconnecting them. A galaxy could be a brain hologram with stars the equivalent of brain cells connected through the dendrite like electric field matrix of interstellar cold plasma. Within our galaxy human brains could have evolved as a fractal of the galactic model.

A key feature of electrically active, non-atomic plasma abounding in space would be its low density. If an atomic nucleus were the size of a golf ball, its closest orbiting electron would be a couple of miles away. If the electron is stripped away to form non-atomic matter then the separation would be far greater so the density of non-atomic matter is much lower than atomic. That is why the density of plasma is less than the best vacuum we can achieve on Earth.

In the past low density matter was called spirit. People didn't know about plasma in space but they believed that space was full of *spirit*. Due to the *scientific enlightenment* based on the material delusion, these traditional ideas have fallen out of fashion. It is possible that in the past people talking about spirits were attempting to explain experiences of life in space. Since ancient times, all races of humankind have held a consensus belief in a universal spirit and lesser spirits that influence life on Earth and the course of human history. The Great Spirit corresponding to God could have been an attempt to describe the universal mind. Spirits, corresponding to the gods, could have been fractals of the universal mind.

The field

Most scientists today treat the idea of spirit with disdain but some are now talking about *the field* which describes electromagnetic fields associated with life. Cold plasma could constitute an electromagnetic field. With the extreme low density of non-atomic cold plasma, this form of matter would be more like an electromagnetic field than body of matter.

An association between electromagnetic fields and life was established before World War II by Harold Saxton Burr, a professor of medicine at Yale. Using a high impedance volt meter Burr measured an electromagnetic field in and around living organisms. This he called the *Life Field* or *L-field*. He published around thirty papers on his extensive research in which he determined that the field had a major impact on the process of differentiation. In Burr's own words:

"When a cook looks at a jelly mould she knows the shape of the jelly she will turn out of it. In much the same way, inspection with instruments of an L-field in its initial stage can reveal the future 'shape' or arrangement of the materials it will mould. When the L-field in a frog's egg, for instance, is examined electrically it is possible to show the

future location of the frog's nervous system because the frog's L-field is a matrix which will determine the form which will develop from the egg." [2]

In her book *The Field* [3], Lynne McTaggart cited a number of credible scientists, associated with universities in the USA and France, Germany and most especially Russia who have investigated electromagnetic fields associated with life. Their research is largely ignored by mainstream science because it conflicts with the tenets of scientific materialism. This is an outstanding example of where the material delusion is impeding the progress of science.

DNA resonance

In my 1990 book *The Vortex: Key to Future Science* [4] (with Peter Hewitt), I predicted a process of *DNA resonance* suggesting how the Life field could underlie the processes of differentiation and evolution. Differentiation is controlled by switching on genes in

2   **Burr**, H.S. *Blueprint for Immortality*, Neville Spearman, 1972
3   **McTaggart** Lynne, *The Field,* Harper & Collins, 2001
4   **Ash** D & **Hewitt** P The Vortex: Key to Future Science, Gateway 1990

DNA that involves complex chemical messaging but the underlying direction of differentiation, *the matrix behind the molecules*, could to be an electromagnetic field. But how would the field influence DNA? I believe the mediating factor may be DNA resonance. The DNA molecule is a double helix coiled upon itself several times. This structure is very reminiscent of a radio or television coil that enables it to resonate with electromagnetic waves.

The idea behind DNA resonance is that information could pass from the life field to the DNA molecule by resonance in much the same way that a program is passed from a broadcast carrier wave to a tuned coil in a radio or television set.

DNA resonance could influence evolution. If the program in the field maintaining the integrity of cells were altered maybe a change could occur in the DNA, or the way the genes are read, to bring about a change in the species. DNA Resonance may explain how evolutionary changes sometimes occur simultaneously in numerous species individuals. Televisions tuned to the same channel respond to the same program on the broadcast signal. So cells tuned to the same species-specific channel of their field would respond to any change in a broadcast program.

Peter Hewitt chose the image of the pastoral god Pan to illustrate DNA resonance. Pan symbolized the field. The pipes of Pan symbolized resonance and his tune symbolized the program in the life field directing the genetic process: *When Pan played the same tune, the species remains the same. When Pan played a different tune the species alters.*

The question is, how could electrical activity in space influence life on Earth? How could non-atomic fields of cold plasma overlay and interact with atomic bodies of matter on Earth to facilitate DNA resonance? This could occur only if the constraints of space and time, electric charge and matter were overcome by space life in the way that people believed was possible for spirit.

Super energy

In the vortex theory space-time, mass and force fields are a consequence of the vortex form of energy. They do not govern energy as energy forms them. Neither can energy be constrained by the speed of light as energy forms light. The speed of light is just the speed of energy in our level of reality. In the vortex theory it is possible for energy to exist with higher intrinsic speeds. Vortices and waves with higher speeds of energy could set up other space time realities. I call energy with higher speeds *super-energy*.

Energy and super-energy could coincide in the same *here and now* as they would be in different planes of space-time; much like layers of paper stacked on a spike. Because slow speeds are subsets of faster speeds, energy is a subset of super-energy. Super-energy could be effective in our space and time because physical space-time would be a subset of the space and time of super-energy. That is possible because space and time are relative to speed of energy.

These exciting possibilities from the vortex theory would allow for a field of super energy to overshadow DNA and mediate by resonance to help maintain the integrity of a cell. This could apply to a multi-cellular organism. If each cell were resonating to a specific program in super-energy field, DNA resonance could help it maintain its differentiated state. Weakening or a disruption of the field could be a cause of disease. Stimulating the field to strengthen it could help the body regain and maintain health. DNA resonance could help provide an account for therapies, including healing, homeopathy, acupuncture and reflexology.

These practices work. The challenge in science has been to understand how they work. The Vortex theory can help us understand how alternative therapies work. In 1986 the British Medical Association published a report on alternative medicine which concluded that while the therapies were effective they could not be endorsed because they were unscientific. The problem is the scientific theory not the alternative medical practice! The consensus belief in the moribund materialistic paradigm proselytised by evangelical pseudoscientific skeptics is a major cause of the prejudice against alternative therapies and the life field. Once scientists distance from the material delusion masquerading as science an unbiased appraisal of life may begin.

Sub-atomic sound

A new understanding of life could begin with treating the longitudinal vibrations of electricity as *subatomic sound*. If life is based on electricity, longitudinal vibrations akin to sound could be fundamental in the transmission of information through space plasma. The silent sound of electricity could be responsible for the expression of creative thought and information transmission throughout the Universe. That would make sense of the opening verse of the Gospel of John: *"In the beginning was the Word and the Word was with God and the Word was God."*[5]

In religion God is equated with light and life, sound and spirit. These are all forms of energy. God could be the conscious expression of mind in invisible non-atomic matter as man is the conscious expression of mind in visible atomic matter. Maybe that is how we are inextricably linked.

I contend we are all expressions of the same creative mind, observed by the same single consciousness, manifest through biological life, on the surface of a spinning orb of atomic matter existing in many dimensions. I believe evolution is influenced, via DNA resonance, by creative intelligence expressed through fields of non-atomic matter. More in the nature of thought than light, these fields of super-energy permeate our world.

---

5   **John** 1:1

Evolution as a Creative Process

I see evolution as a creative process that is as much art as science. In my view our world is both canvas and laboratory to intelligent, non-biological life forms we have yet to comprehend. I also see biological life resonating to the tune of plasmic non-biological life that forms living templates of subatomic sound beyond the reach of our limited senses.

So who or what are these mysterious life forms from the depths of space that set in their own fields of life the templates for life on Earth, even the human body? I believe they are us. In my view we are not our bodies. I believe we are ancient, non-biological life forms, living fields from electric worlds of non-atomic matter, responsible for the evolution of biological life.

We as collective consciousness have monitored the guiding intelligence behind evolution from single cell to multicellular organisms. Collectively we are the observer and the expression of creative imagination that has engineered watery bodies so we can inhabit both plant and animal, enabling the universal consciousness that we are to experience a very rare world of atomic matter.

Our current peak adventure is human life. That too will pass and then where will we go and what will we do? That is for us to imagine! Meanwhile in the eternity of now there is no where else to go and nothing better to do than to be fully present in the individual mask we occupy until in the fullness of time we surrender to the real world, the oneness of being, the infinite continuum of life.

Appendix 1: Mass

Defining Energy and Mass

To appreciate vortex energy as mass consider Einstein's equation $E=mc^2$. If 'E' quantifies energy and the symbol 'c' denotes what energy is, then a complete definition of energy would be 'Ec'. To understand this imagine going into a shop and asking for three. The shopkeeper could not serve you because he wouldn't know whether you wanted three apples or three oranges. If you went in and asked for apples he couldn't serve you until you said how many apples you want. He could only serve you when if you ask for three apples. To define anything it is necessary to stipulate how much and what an item is. 'E' tells us how much energy we have. That is like the number three in the analogy. 'c' tells us what energy is – the speed of light. That is like the word apples in the analogy. 'Ec' defining energy is like asking for 'three apples'. If this complete definition for energy is applied to Einstein's formula $E=mc^2$, then the equation expands by the constant 'c' to $Ec=mc^3$. This expansion of the equation describes mass as the distribution of energy 'c^3' in three dimensions.

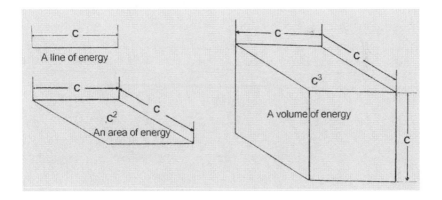

'c' describes a line of the movement of in one dimension. 'c^2' representing an area describes the movement of light in two dimensions. 'c^3' representing a volume describes the movement of light in three dimensions. The expanded equation $Ec=mc^3$ depicts mass as a three dimensional form of movement of light. Three dimensional extension is a fundamental property of matter. Common experience reveals every massive object to have three dimensional extension. Most people accept this without question but when energy is described as a line of the movement of light, Einstein's famous equation reveals why mass extends in three dimensions. In the vortex theory mass is defined as quantity of vortex energy. Mass is a property of the vortex, not the energy within the vortex. Photons do not possess mass because they are wave, not vortex forms of motion.

The law of Conservation of Angular Momentum

Momentum is defined as mass x velocity. Momentum laws applying to vortex particles in matter cannot be applied to the energy within the vortices. For example, the law of conservation of angular momentum can be applied to the motion of vortices because the vortices have mass but the law cannot be applied to the energy that forms the vortex because momentum incorporates mass and energy has no mass.

The Inertia of mass explains away materiality

The vortex form of energy gives rise to the illusion of materiality in our world. The illusion of material substance originates from the 'inertia' of mass. Inertia is a fundamental property of matter. This defining property of 'materiality' is explained away by the vortex model.

Mass and Potential Energy

The vortex model accounts for potential energy; that is it explains how energy is locked up in mass and the forces associated with mass. A straight line vector represents the maximum velocity of any line of movement. If the line begins to follow a curved

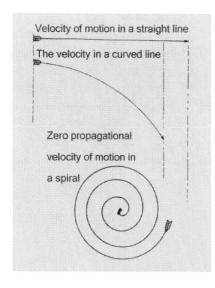

path the straight line velocity vector would diminish. If the line of motion then spins in a spiral path its velocity in the straight line – that is its propagation vector – would be zero. So the spiral motion of energy would set up a system of apparent stasis. This is how vortex motion would set up the system of passive inertia we recognise as mass and potential energy.

The concept of inertia is important for describing fundamental particles as forms of energy. Because energy is neither created nor destroyed vortices of energy must be infinite in extension. Therefore the concept of size would not apply to a subatomic vortex. A vortex particle can only be defined by its mass; its potential energy and passive inertia.

Appreciating Models

The wave and vortex serve as models for energy. They are equivalent to pictures of trees in a school. The pictures are images of trees. They are not real trees but they enable children to learn about trees so they can appreciate the real things when they go into the woods. Likewise scientific models are not real; they just help us appreciate reality. For example inertia is a reality represented by the wave and vortex models. The models enable us to see that inertia is a fundamental property of energy arising from the dynamic state. In the models we see energy in a state of action and energy in a state of apparent rest setting up active and passive inertia. The models help to clarify our understanding of inertia. Without the vortex model we don't understand inertia.

As Richard Feynman said *"...the laws of inertia have no known origin."*[1] We think of inertia in terms of the static state which is bewildering because if everything is energy how can anything be static! The vortex model enables us to appreciate how static inertia originates from the dynamic state of energy in spin and the wave model helps us see how inertia exists in the kinetic wave form state.

1   **Feynman** R. *The Character of Physical Law*, Penguin Books 1997

Appendix 2: The Neutron

Discovery of the Neutron

The neutron is particle without charge in the atomic nucleus. It was discovered in 1932 by the English physicist James Chadwick at the Cavendish Laboratory in Cambridge for which he received the Nobel Prize in 1935.

Structure of the Neutron

The chargeless neutron is formed when negatively charged electrons are captured by the atomic nucleus from the innermost K-orbit of an atom. When this occurs a positively charged proton disappears. The neutron has the sum mass of an electron and a proton and as it is neutral in charge it would be natural to assume it is a bound state of an electron and proton with each canceling out the charge of the opposite. This occurs in the atom which is electrically neutral because it contains an equal number of opposite charges. This obvious conclusion is reinforced by Beta decay in natural radioactivity. Outside the nucleus of an atom neutrons decay in a matter of minutes into an electron and a proton. [1]

Electric dipole of the Neutron

In 1957, Smith, Purcell and Ramsey discovered that a neutron has a slight electric dipole moment, suggesting it is not an entirely neutral particle. On one spot the neutron displays a minute negative charge, in the order of a billion, trillion times weaker than that of a single electron. This suggests that the neutron is a bound state of opposite charges which mostly cancel each other

[1]   **Richards, et al**, Modern Univ. Physics, Addison Wesley 1973

out. Proponents of the standard theory, opposed to the obvious bound-state, latch onto the large margin of error on the electric dipole moment of the neutron - equal to the charge on an electron x10^{-20} (-0.1 +/- 2.4) – Emilio Segre said in his book 'Nuclei and Particles', "*... this moment could be exactly zero, in agreement with the theory.*" [2]

The large margin of error due to the extreme weakness of the measure of charge does not allow the 1957 experiment to be taken as conclusive but neither does it allow the measure to be taken as zero; unless the objective is to dismiss an inconvenient truth.

Magnetic moment of the Neutron

The presence of charged particles in the neutron is supported by the magnetic moment of a neutron at 1.91 nuclear magnetrons.[1] The neutron, if it were a truly neutral particle, would have no magnetic moment because the magnetic moment of a particle is created by the spin of its charge. The magnetism of a neutron adds support to the view that it is a bound state of two opposite charges, which mostly cancel each other out, not a single particle with no charge at all.

Directional emission of Beta decay electrons

Chien Shiung Wu observed when the nuclei of radioactive atoms are aligned in a magnetic field they emit more beta electrons in one direction more than in another.[3] This suggests that in radioactive decay electrons emerge from specific sites in the atomic nuclei which is suggestive that neutrons in the nuclei have electrons bound within them. This experiment supports the bound state model for the neutron. The formation of strange particles around electrons after high energy bombardment of atomic nu-

2   **Segrè** Emilio, *Nuclei & Particles* Benjamin Inc 1964
3   **Richards, et al**, Modern Univ. Physics, Addison Wesley 1973

clei adds further support to electrons being at rest in the nucleus of an atom.

Quantum spin of the Neutron

An argument against the bound state theory is that the quantum spin of a neutron is the same as a proton or electron. Physicists argued that a neutron cannot have the same value of quantum spin as a proton or an electron if it is a combined state of them both. However, if a light electron were bound by a massive proton, its quantum spin would be effectively hidden. The enormous passive inertia of the proton compared to an electron, conferred by its greater mass, would account for the neutron appearing with the spin of the proton. The law of conservation of angular momentum does not allow for the quantum spin of an electron to be lost in a neutron but physics does allow for conservation laws to be upheld in the formation and decay of unstable particles – so long as nothing is gained or lost in the overall process – and quantum spin is conserved in the overall process of formation and decay of the unstable neutron.

Mass of the Neutron

The rest mass of an electron is 0.911×10^{-30} kg. The rest mass of a proton is 1672.62×10^{-30} kg. The rest mass of a neutron is 1674.92×10^{-30} kg; so it has a mass of 1.389×10^{-30} kg in excess of the sum mass of an electron and proton. This is equal to 1.5x the rest mass of an electron. In the neutron, an electron could not possess a momentum in excess of that allowed by the energy locked up in 1.389×10^{-30} kg of mass by $E=mc^2$. [4] The certainty in the mass of a neutron points to considerable uncertainty of the uncertainty principle. All the evidence about the neutron stacked against it supports Einstein's position that Heisenberg's principle is flawed and the quantum mechanics

4   **Richards, et al**, Modern Univ. Physics, Addison Wesley
      1973

derived from it is unsound. Skeptics deride people who believe in fairies but the physics they believe in is no less a fable!

Failure of the uncertainty principle with the Neutron

The Heisenberg uncertainty formula fails with the as an electron-proton bound state. If an electron was bound in a the certainty of its position would lead to an uncertainty in its momentum reflected in a momentum range between zero to velocities approaching 99.97% that of light. Electrons with that order of velocity would have a mass 40x that of an electron at rest. The mass of the would then range from 1673.5×10^{-30} kg to 1710×10^{-30} kg. However, all s possess the same mass of 1674.92×10^{-30} kg., which indicates there is no detectable, indeterminacy in the momentum of the electrons that numerous experiments show they contain.

Vortex theory for the proton

To understand the imagine an electron vortex falling into a proton where it becomes enmeshed and held by the force of electric charge acting between them. In the , the proton would immobilize the electron by virtue of its greater static inertia. Unable to undergo annihilation, because of the greater mass of the proton, and meeting the resistance of swirling captured energy within the proton vortex, the electron would come to rest just outside the region occupied by the meson energy. It would lodge on the proton 'surface' delineated by the range of the strong nuclear force. The kinetic energy of the electron – caused by its partially captured energy equivalent to 1.5x its mass – would act against the force of attraction binding the electron to the proton. Held to the proton by opposite charge attraction, but driven by its kinetic inertia, one could imagine the electron pulling away from the proton in a constant struggle to break free of the . This would account for the instability of the .

instability within atomic nuclei

With the high kinetic energy of the bound electron it is unlikely that neutrons are stable within atomic nuclei. It is more likely that there is a constant interchange of electrons between protons and neutrons within the nucleus.

Cat and mouse analogy for the Neutron

To understand the exchange of electrons between protons and neutrons in an atomic nucleus imagine the electrons as mice caught by a group of cats. The proton is represented by a cat without a mouse and the neutron as a cat with a mouse. As fast as a mouse escaped from one cat in the group another would catch it.

The cat and mouse analogy for radioactivity

The spontaneous explosion of the high-energy electron from an atomic nucleus is the radioactivity called beta-decay. The loss of electrons from neutrons is the source of beta-decay but the reason why some atoms are radioactive and some are not can be understood from the cat and mouse analogy. The greatest chance of a mouse escaping from a group of cats would be when the group is either small or large. In a large group of cats, where there are many mice involved in the exchange between cats, the chance of a mouse escaping would be greater than in a medium size group. Thus it is that the chance of electron escape from atomic nuclei would increases with size. That accounts for why the heavy atoms tend to be radioactive. My father researched the ratio of neutrons to protons in atoms and discovered that although it didn't apply in all cases there was a tendency overall for an exceptionally high ratio of neutrons to protons to be associated with natural radioactivity. This is represented by a higher ratio of mice to cats in the analogy leading to a greater statistical opportunity for escape. In a small group of cats there would also be a chance for the occasional mouse to escape. Electrons sometimes escape from small atomic nuclei where again there is a high ratio of neutrons to protons.

The Neutrino

In the 1920's, a serious problem arose in the study of beta-decay. Physicists found that the electrons, emitted from radioactive atoms, emerged with a wide range of different energies, whereas the total available energy released from the neutron decay was the same on every occasion. The problem was trying to account for the energy lost from the nucleus when an electron emerged with less than the standard energy. Careful studies showed that the surplus energy was not appearing as heat, light or gamma rays. In 1930, Wolfgang Pauli suggested that another particle was released in the process of beta decay, which carried away the excess energy. Enrico Fermi dubbed it the neutrino meaning 'little neutral one'. Neutrinos are very nebulous. They can pass right through the Earth without reacting with matter at all. Neutrinos have neither mass nor charge so they cannot be vortex forms of energy. If they carry energy away in beta decay they must be a form of radiant energy. The photon is a form of radiant energy without mass so the neutrino would appear to be more akin to a photon of light than a particle of matter. The neutrino has a quantum value – quantum spin – denoted by a half Planck's constant which suggests that it might be half a quantum. One possible account for the neutrino is that it is a single line of the movement of light in the wave form.

Cat and mouse analogy for neutrinos

In the cat and mouse analogy neutrinos would correspond to all mice having the same length of tail but some mice losing a part of their tails as they escape from the cats. The fragment of tail lost to the claws and teeth of the cat would differ with each escaping mouse. That would correspond to the different amounts of energy unaccounted for in beta-decay. The intact tails would correspond to the total available energy for beta-decay which is always the same. The fragments of tails lost in the fray would correspond to neutrinos. Treating neutrinos as an aspect of radioactivity that defies measure enables the postulate that they are invisible, single string wave-train particles known only because they carry energy away from the total available in beta-decay. That can help in the understanding of wave train forms of energy and the way they interact with matter.

Appendix 3: Wave-particle duality

Wave theory for light

The wave nature of light was established by the English scientist Thomas Young, in an experiment called Young's slits experiment.

This experiment appeared to contradict Newton's particle theory for light but Einstein re-established the particle theory alongside the wave theory. The wave-particle conjunction came to be known as 'wave-particle duality'.

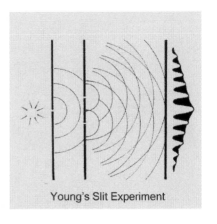

Young's Slit Experiment

Wave theory for electrons

In 1922, the French physicist, Louis de Broglie turned the dilemma round by suggesting that particles of matter could be treated as waves. His idea was confirmed in 1927, when two American scientists at the Bell Telephone Laboratory, C.J. Davisson and L. H. Germer, discovered that a beam of electrons created the same characteristic wave patterns as a beam of light in Young's experiment. They demonstrated that electrons

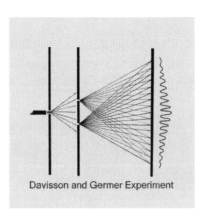

Davisson and Germer Experiment

have the same diffraction patterns as those predicted by Bragg for X-Rays. Davisson and Germer received the 1937 Nobel Prize for physics for confirming Broglie's hypothesis.

Erwin Schrödinger, an Austrian physicist, embraced de Broglie's concept when in 1926, he developed a wave equation which could be applied to the motion of the electron in a hydrogen atom. He received the Nobel Prize for physics in 1933.

The Principle of Complimentarity

Particles of light and matter never behave as both particles and waves in the same experiment. In one experiment they appear as particles and in another they behave as waves. Bohr expressed this as the 'Principle of Complimentarity'. Electrons and photons move as waves but are generated and absorbed as particles. An experiment, such as interference or diffraction, shows the effects of particles of energy when they are moving. Here the wave properties are revealed. In an experiment, such as the photo-electric effect or Compton scattering, the motion is wholly or partially absorbed. Here the particulate nature is revealed. Something cannot be simultaneously moving and not moving. Therefore it is obvious that no experiment can show both particulate and wave properties of a particle of energy. The principle of complimentarity supports the vortex theory of energy existing with two complimentary inerties; activity and passivity. Some experiments demonstrate the active inertia of the wave form of energy and others the passive inertia of the vortex form.

Appendix 4: Electromagnetism

Richard Feynman's theory for electromagnetism

Richard Feynman commented: *"It is simple therefore it is beautiful."* The theory he proposed for electromagnetic fields was elegant. Wave particle duality, the equivalence of mass and energy and Maxwell's electromagnetic theory for light had led to the belief that matter and light can be unified in a single field theory. Then Maxwell's concept, that light consisted of electric and magnetic fields, was reversed by Feynman. He suggested that the electric and magnetic fields of force associated with matter were caused by light. Feynman's concept was that *virtual* particles of light pop up alongside charged particles to set up the electromagnetic forces. After an exchange with another charged particle they vanish again. The whole process of the light appearing and disappearing was considered to occur within the bounds of Heisenberg's uncertainty principle so the particles of light could never be detected as such therefore they were called 'virtual photons'. One could imagine broadsides of virtual photons being fired between charged particles as though they were minute battleships. These energetic interactions were thought by Feynman to be responsible for pushing them apart or pulling them together.

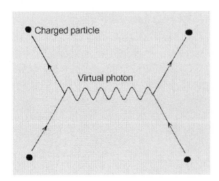

Feynman Diagrams

Feynman designed diagrams to illustrate the exchange of virtual photons.

These Feynman Diagrams have also been used to illustrate the exchange of other force-carrying particles to explain the action of other forces in the theories of quantum mechanics.[1]

The vortex theory overcomes particle mass problems

The neutron-proton mass problem does not occur in the vortex theory. Because electric charge is an expression of the dynamic nature of the subatomic vortex of energy there is no requirement for subatomic particles to have any additional energy to account for their charge interactions. It is the electron bound to a proton, causing neutrality of the neutron that explains the neutron's greater mass. Also in the vortex theory there are no problems associated with locality of forces because vortices of energy are infinite extensions. That is why the forces of electric charge, magnetism and gravity act over infinitely large distances. As the quantum vortex extends in three dimensions the drop in intensity of vortex energy would obey the 'Inverse Square Law' which applies to all three dimensional extensions.

The Inverse Square law

The inverse square law directs that: *The intensity of vortex energy at a distance from the center of the vortex would be inversely proportional to the square of that distance.*

This law indicates that as the distance from the vortex centre doubles, the intensity of vortex energy drops to a quarter of its original value. As the distance trebles the intensity drops to a ninth. The law also shows that with distance, while the intensity of energy rapidly diminishes it never falls to zero. The fraction may become small, even infinitesimal, but it never ceases to be. This infinite extension of the quantum vortex suggests that every electron occupies the entire universe. It could provide a very

---

1   **Calder** Nigel, *Key to the Universe,* BBC Publications

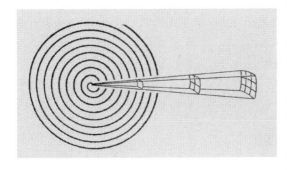

simple account for the non-locality of electrons and Einstein's EPR, 'spooky action at a distance' paradox; the way subatomic particles affect each other, instantaneously, over infinite distances.

Coulomb's law

The Inverse Square law applied to vortices of energy accounts for Coulomb's Law. In the eighteenth century the French scientist, Charles Coulomb first observed that the force of interaction between charges is inversely proportional to the square of the distance between them.

Opposite signs of charge

In the spherical vortex there are only two ways in which energy can flow – into or out of the centre. This property of the quantum vortex would account for the opposite signs of charge, 'positive and negative'. Positive charge is represented by the flow of energy out of the centre of the quantum vortex and negative charge by the flow of energy into the centre.

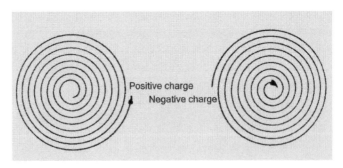

Positive charge
Negative charge

Charge is independent of mass

When electric charge is treated as a vortex interaction, each vortex of energy would represent an indivisible unit of interaction. It would be an irreducible quantum of electric charge therefore charged particles of matter, as vortices of energy, would all have the same unitary electric charge regardless of their mass. Electric charge would build up as an integer multiple of single units of charge; each unit representing the action of a single vortex of energy. This is precisely how electric charge accumulates – by the addition of individual charged particles. To appreciate this point, think of an army. An army is made up of individual soldiers and changes in size by reinforcements or losses of whole numbers of men. It doesn't matter whether the soldiers are big or small; each man carries a rifle to enable him to act from a distance, therefore his effectiveness is independent of his mass. Elementary particles have never been observed with a value of charge greater or lesser than unity, which fits with the vortex theory. A soldier would not be effective with half a rifle!

Fractional charge assumption in the Quark theory

All charged particles have unitary charge, either positive or negative, which is independent of their mass. Charges add up or cancel out by the accumulation of particles with the same or opposite whole unit of charge, yet in quark theory there is an arbitrary assumption of fractional charge! Up-quarks are supposed to have 2/3 charge and down-quarks 1/3 charge. In the proton 2/3 + 2/3 - 1/3 = 1 which gives the proton unitary charge. In the neutron 1/3 + 1/3 - 2/3 = 0 which gives the neutron zero charge.[1] Physicists and skeptics swallow this nonsense. They don't question the groundless speculations in quark theory. The fact is there is no evidence for fractional charge in nature.

Dual model assumption in the Vortex theory

To account for electric charge it is necessary to assume two models for the vortex. In the distant action of electric charge, the spherical vortex behaves like a system of discontinuous concentric spheres whereas close to the centre, the vortex appears to be a continuous spiral. Both the continuous and the discontinuous models apply to the quantum vortex.

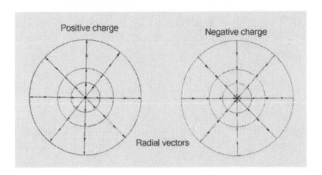

Dual model account in the Vortex theory

The dual nature for the vortex can be explained by the ball of wool model. With a wool ball the spiral would be tighter and more pronounced near the centre. As the wool ball increases in

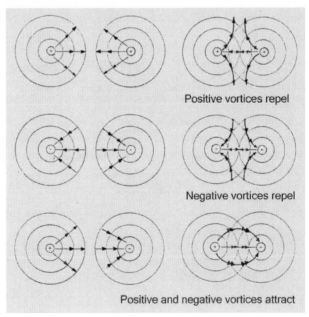

Positive vortices repel

Negative vortices repel

Positive and negative vortices attract

size, the tendency to be a spiral decreases and tendency to be concentric spheres increases. At a distance from the centre the flow of energy in the spherical vortex would take the form of concentric spheres of energy either growing out of or shrinking into the centre. The atom is mainly space. In the distance be-

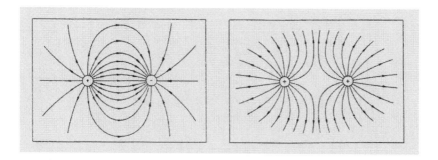

tween the nucleus and the orbiting electrons, vortex interactions would tend to be more the effects of expanding or contracting concentric spheres than vortex spiral interactions. This appears to be the case with electric charge. Electric charges appear as interactions between growing or shrinking spheres of energy and these motions are effective along their radii, therefore electric charge can be described as radial effects of vortex energy or 'radial vectors' The word vector is used to denote the direction in which a motion is effective.

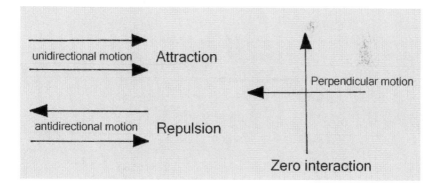

Radial vectors of charged particles

The French scientist, Charles du Fey recognised that like charges repel and unlike charges attract. These characteristic interactions of charges can be derived by plotting the radial vectors of overlapping concentric spheres of vortex energy.

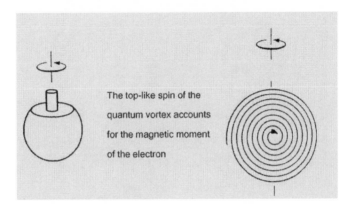

The top-like spin of the quantum vortex accounts for the magnetic moment of the electron

Lines of electric force

The points of interaction between overlapping radial vectors of vortex energy can be used to plot out the lines of force acting between charged particles. These correspond to the textbook diagrams of the characteristic electric lines of force acting between charged particles.

Vector laws for attraction, repulsion and no interaction

If the flow of energy in subatomic vortices is opposite in direction they act against each other and repel. If the flow is in the same direction they act with each other and attract. This is why particles with the same charge repel and particles with opposite charge attract. Mid-way between attraction and repulsion is zero interaction and mid-way between movement in the same and opposite direction is movement at right angles. These lead to the vector laws for vortex interactions.

Magnetism

Perpendicular vectors set up the distinction between electric and magnetic fields. Charged particles set up magnetism if they move at right angles to the charge. The top like spin of an electron in the atom gives rise to its magnetic moment.

The magnetic moment of an electron results from its rotation. This is perpendicular to the vortex of energy that forms it and its field of electric charge. The two types of spin set up two distinct force fields; charge and magnetism.

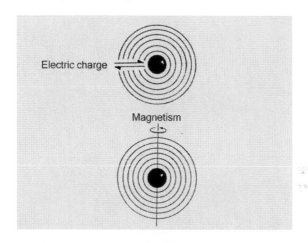

Tangential vectors to charge set up magnetism

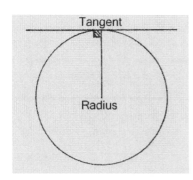

If charge is the growing or shrinking of the concentric spheres of energy forming a subatomic vortex, the secondary 'rotational spin' of the concentric spheres causing magnetism will be effective at a tangent to them. Because a tangent is at right angles to the radius of a sphere, the rotation of vortex energy responsible for magnetism is effective at right angles to the flow of vortex energy causing its charge. Tangential vectors to the concentric spheres of charge set up magnetism.

Imagine a party of children in a park. Some are blowing balloons and some are playing on a roundabout. Close your eyes and visualise the scene. The motion of the expanding balloons is perpendicular to the turn of the roundabout so it is that the primary in or out 'vortex' spin forming the subatomic particles is perpendicular to their secondary rotational or 'roundabout' spin.

Perpendicular vectors in photons of light
Vectors of energy set up force fields. Perpendicular vectors of energy create distinct fields because if the fields are perpendicular they do not interact with each other. This helps us understand the photon of light. Each photon consists of two vectors of energy undulating at right angles. The lack of interaction between these perpendicular fields of energy would explain the lack of interference between them.

Vector interactions between light and matter
Perpendicular vectors of energy in the photon interact with the perpendicular vectors extending from subatomic particles of matter in kinetics. The lack of distinction between energy vectors in matter and light, have led, in part, to the attempts to include matter and light in a single unified field theory. (Light here is taken to include heat, radio and gamma rays.)

Magnetic and non-magnetic atoms

In the atom, electrons occur in pairs. Each member of the pair spins in the opposite direction. Because the relative directions of these motions are parallel and not perpendicular they interact and because the interactions are opposite they cancel each other out. The result is a cancellation of the magnetism. This is why most atoms are non-magnetic. Natural magnetism arises from atoms containing an unpaired electron. Because this magnetism is not cancelled it confers magnetism on the atom as a whole. Atoms of iron are magnetic because of the spin of an odd electron but iron only becomes magnetic when the atoms are aligned in the same direction. If they lose their alignment, through heating or hammering, the iron loses its magnetism.

Whirlpool and top analogy for charge and magnetism

The difference between electric charge and magnetism can be appreciated by analogy with whirlpools and tops. The spin of water in a whirlpool represents charge and the spin of a top represents magnetism. Now imagine a whirlpool rotating in a tidal current. This illustrates an additional motion imposed on the vortex that sets up an additional field of force.

Superimposed vectors of charge and magnetism

When nothing exists but motion all forms and forces in the Universe consist of vectors of motion interacting with each other and some of the interactions can be quite complex as one vector of motion superimposes on another. The radial flow of concentric spheres of energy is the primary motion in the vortex. This movement gives rise to the radial vectors of electric charge. The vortex can then rotate. This secondary spin is a superimposed motion, a perpendicular vector that gives rise to the natural magnetism, the magnetic moment of a particle. The rotating vortex can then vibrate or move. These tertiary motions superimposed upon the primary and secondary movements of a subatomic vortex contribute to the force fields. The type of field depends on the relative directions of the overlapping and interacting energy vectors. The interactions of vortex vectors follow self evident rules.

The rules of vortex vectors

If superimposed vortex vectors move relative to one another along the radii of the vortices they interact as an electric field. If their relative movements are at a tangent to the vortex, they interact as a magnetic field.

The interaction rules of vortex vectors

The interaction rules between vortex vectors are simple and obvious. When vortex vectors move with each other in the same direction they attract. When they move against each other in the opposite direction they repel.

The interaction rules between electric charges

The vortex rules of interaction are demonstrated by electric charges. With positive charges the concentric spheres of energy expand. When they overlap they move against each other in opposite direction as they grow out from their vortex centres. With negative charges the concentric spheres of energy contract. When they overlap they move against each other in opposite direction as they shrink into their vortex centres. The vortex vectors of like charges are relatively effective in opposite direction so they repel. When positive and negative charges overlap one set of vortex vectors are directed inwards while the other set are directed outwards. The relative vectors of the opposite charges are effective in like direction so they attract.

The interaction rules between magnetic moments

The interaction rules apply to magnetism as well as charge. There is a magnetic repulsion between two electrons spinning in same direction. This is because the rotating vortex vectors are in the opposite direction where they meet.

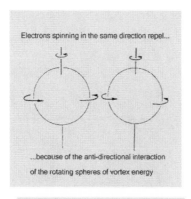

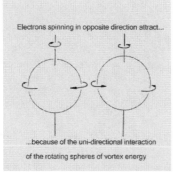

There is a magnetic attraction between two electrons spinning in opposite direction. This is because the rotating vortex vectors are in the same direction where they meet.

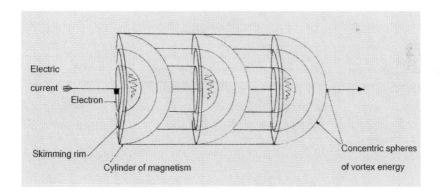

The magnetism around a current of electricity

When electron vortices flow in an electric current there are vortex vectors associated with this motion that are effective at a tangent to them. These set up concentric 'rims of magnetism' that surround each moving vortex at right angles to the current of electricity. The concentric rims of magnetism would add up to form concentric cylinders of magnetism. Imagine slicing an onion then skewering it through its centre. In the line of the skewer, representing the current, the moving onion rings would circumscribe concentric cylinders.

Vortex vector rules for electric currents

If two current carrying conductors are laid side by side they will

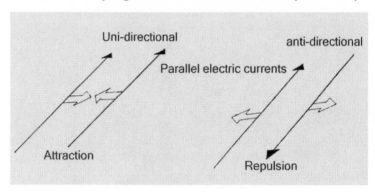

attract if the current is flowing in the same direction, but repel if it is flowing in the opposite direction. This effect was first observed by André Ampere.

When the concentric spheres of vortex energy, extending from the two parallel electric currents, move against each other in opposite direction they repel. When they move with each other in the same direction they attract. The vectors are effective at a tangent to the concentric spheres of vortex energy so the interactions between the current carrying conductors are magnetic.

Interactions between radial and tangential vectors

Michael Faraday discovered that a stationary magnetic field would not interact with a stationary electric field. Radial vortex vectors set up electric fields. Tangential vortex vectors set up magnetic fields. These fields are perpendicular therefore they are distinct and non-interactive. Faraday then found that a moving magnetic field will interact with an electric field. The 'movement' of the magnetic field is a tertiary motion superimposing on the primary motion of the vortex causing charge and the secondary motion causing magnetism. This new motion will set up a new set of vortex vectors. If they are radial in effect they will interacted as an electric field. If they are tangential in effect they will act as a magnetic field.

Faraday discovered that a moving magnetic field will act as an electric field and induce a flow of electricity in a wire. He used this to invent a rudimentary dynamo. This discovery is the basis of electricity generators in power stations, motor vehicles and wind turbines.

Faraday also discovered that a growing or decaying magnetic field will interact with an electric field. In a growing or decaying magnetic field the intensity of vortex energy is changing. The intensity of energy in the spherical vortex is uniform over the surface of each concentric sphere whereas it varies along the radius. Uniform movement at a tangent to these spheres would not incur any change in intensity of energy. This is the characteristic of a magnetic field. However, movements along the radius of the vortex are associated with change in intensity of energy. This is the characteristic of an electric field.

Any change in intensity of vortex energy would be effective as an electric field and would interact with another electric field. A growing or decaying magnetic field will interact with an electric field because it is an 'electric' effect. Alternating currents of electricity grow and decay many times a second. The growing and decaying fields of magnetism around them are used to operate transformers that change voltages from those in the national grid to the charger for a laptop computer.

Effect of antimatter on vortex vectors

Where nothing exists in quantum reality but motion the things that matter are vector interactions; the relative directions and intensities of interaction between different forms of motion. The primary direction of vortex motion is paramount. If the direction of primary motion in a subatomic vortex of energy is reversed, then this would affect any secondary motion superimposed upon it.

Particles with the same sign of charge, but spinning in opposite direction, experience a magnetic attraction. If the sign of charge of one of them is reversed then the magnetism will reverse and set up a force of repulsion. This effect has been observed between particles of matter and antimatter. Just prior to annihilation electrons and positrons, undertake a 'death-dance' called 'positronium'. If the electron and positron are rotating in the same direction they attract each other and consequently annihilate sooner. If electron and positron rotate in the opposite direction the magnetic repulsion delays their annihilation. This shows that reversing charge has the effect of reversing the direction of magnetic interaction.

Electricity

In the vortex theory there is an account for electricity. When electrons are free to move in a conductive substance, they can be aligned by the application of a potential gradient along the conductor. Once they are aligned, movement can be transmitted as vibration, from one electron to another, down the voltage gradient. Electricity is the result of a longitudinal – compression and rarefaction - vibrations of electrons in matter.

A parallel can be drawn between electricity and sound as sound is formed through longitudinal vibrations in matter. To understand electricity, imagine electron vortices shunting against one other. Because of the infinite extension of vortex energy, the shunt or longitudinal vibration would pass from one electron to another without them ever coming into direct contact. The shunt or longitudinal vibration is a radial vortex vector. Electricity is the passage of the shunt down the line of electrons rather than the passage of the electrons themselves. While the electrons drift slowly in the potential gradient – approximately three hundred kilometers per hour – the electricity travels at approximately the speed of light – three hundred thousand kilometers per second.

The parallel between electricity and sound makes this clear. Sound passes through air at about 1,000 miles per hour whilst, even in a hurricane, the air itself moves at a tenth of this speed. Sound isn't the flow of the air; it is a vibration passing through the air. Sound is a compression and rarefaction that is tantamount to a shunting of air molecules. Like sound, electricity is the passage of activity through a medium rather than the passage of the medium itself. Sound is the vibration of atomic matter. Electricity is the vibration of subatomic matter. Electricity could be thought of as subatomic sound.

Appendix 5: The Quantum

Planck's Constant

Max Planck proposed the energy 'E' of a quantum to be proportional to its frequency 'v'. The constant of proportionality came to be known as Planck's constant 'h'.Because Planck was concerned with black body radiation his constant defined the parameters a quantum of radiant energy.

Fermions and Bosons

The energy in many subatomic particles is denoted by half rather than the whole of Planck's constant so particles with the quantum value ½h are dubbed 'fermions' after the physicist, Enrico Fermi. It the particle has a quantum value h it is called 'boson' after the physicist, Satyendra Nat Bose.

Quantum Spin

The term 'quantum spin' is used to describe the quantum designation of particles; whether they are fermions of bosons. No one is sure what quantum spin is but one thing is for sure, its current use blocks the description of the subatomic vortex as quantum spin; which is what it is. Because the word spin is used to denote tangential as well as radial vectors of spin the description quantum vortex is more definitive for subatomic particles.

The Quantum is not fundamental

Originally the atom was thought to be the most fundamental particle of matter then later it was discovered to be a complex form of subatomic particles. History appears to have repeated itself with the quantum. Now it seems the quantum is not a fundamental form of energy. Later discoveries can add confusion to physics especially to the use of terms like atom and quantum. With advances in physics there is a continual need to revise nomenclature and redefine terms in line with new discoveries and changing theories.

The Neutrino

The neutrino would appear to be more fundamental than a quantum because it has a quantum value of half Planck's constant i.e. ½h. Accounting for energy lost during the breakdown of a neutron, the neutrino appears to be a radiant form of energy. Unlike the particles of radiant energy in heat and light, gamma rays and radio waves, the neutrino has a very low level of interaction with matter. Its presence in the neutron suggests it does interact with sub-atomic particles.

The Quantum and the Neutrino

The electromagnetic theory for light presents the photon or quantum as a bound state of two wave-trains of energy each propagating at right-angles to the other, which figures with its being a boson. It seems that the boson quantum delineated by 'h' is not the most basic unit of energy. A single field of energy delineated by ½h appears to be more fundamental If the quantum delineated by 'h' consists of two wave vectors of energy – lines of the movement of light – undulating at right angles then the neutrino delineated by ½h could be half a quantum that is a single wave vector of energy; a single undulating line of the movement of light. If the neutrino, as a single wave form line of the movement of light, does not interact with atomic matter that would suggest the perpendicular arrangement of two undulating lines of light is key to interactions with the electromagnetic force fields associated with atoms

Quantum interactions with atoms

The atom is mainly space, occupied by fields of electric charge, magnetism and gravity. Sub-atomic particles, and atoms, interact with each other at a distance by means of these forces. It is the energetic interaction between photons of heat and light and atoms that we witness on a day to day basis. It could be the perpendicular wave train form of the photon, matching the perpendicular electromagnetic fields extending from subatomic particles, that enables it to interact with the fields in atoms. These interactions have contributed to the mistaken assumption that forces associated with subatomic particles of matter are the same as the electromagnetism in light. That led to the speculations in quantum electro dynamics of virtual photons as the cause of the electromagnetism in matter.

Only one field in the photon interacts with matter

In experiments such as fluorescence and polarization it seems only one of the fields in the photon actually reacts with matter. This is also evident from photography. Only one field of energy in light reacts with the photographic plate. Only one wave vector in the quantum interacting with matter suggests the perpendicular arrangement of two fields inn light may not be important in wave-vortex interactions. The question is, what role does the wave train in the photon of light have? Why does it not engage with the electromagnetic fields in the atom? Is the non interactive wave train linked in some way to the neutrino?

Quantum interactions with subatomic particles

When it comes to quantum interactions with subatomic particles it is hard to imagine the complex double field photon driving into a single subatomic vortex. The image of a single field of energy driving into the vortex is confirmed by quantum mechanics where the kinetic energy of elementary particles appears to be gained or lost in integer multiples of ½h.

Money analogy for quantum interactions

Quantum interactions can be understood by analogy with money. In the world at large, mechanical events involve quantities of energy that are comparable to multi-billion dollar transactions. Quantum theory, concerned with the kinetics of elementary particles, deals in energy quantities comparable to cents. If the universal cent is a line of the movement of light denoted by ½h it makes sense that quantum interactions are accounted in integer multiples of ½h. If this model is correct it confirms that single lines of the movement of light in wave form do interact with atomic matter and the neutrino remains an enigma.

Quantum interactions and antimatter

A question remains in quantum interactions. If only half a quantum of wave energy appears to interact with a vortex particle of matter, what happens to the other half? The answer may lie with antimatter. And antimatter may resolve the enigma of the neutrino. It may be that the neutrino is a quantum in reverse where the wave that is supposed to be in the world of matter has flipped into the world of antimatter and vice versa. If that were so, if matter did interact with neutrinos we could go back in time.

Appendix 6: Antimatter

History of antimatter

Antimatter was predicted in Victorian physics. Karl Pearson proposed 'squirts' forming sources in matter and sinks in negative matter and Arthur Shuster wrote to *Nature* in 1898 using the term 'antimatter' for the first time. Shuster hypothesized anti-atoms, antimatter systems and discussed the idea of matter and antimatter annihilating. Antimatter was again predicted in 1928 by Paul Dirac. It all seemed like science fiction until in 1932 Carl Anderson at the California Institute of Technology spotted a positively charged electron emerge from lead bombarded with gamma rays. He called the new particle a 'positron'.[1]

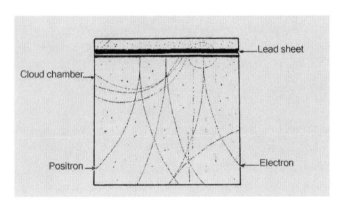

1   **Richards et al,** Modern University Physics, Addison Wesley 1973

Pair Particle Production

Anderson discovered positrons were created with electrons in pairs. He realised only when gamma ray photons had energy exceeding the mass energy equivalent of an electron and positron were the pair particles produced; so the production was from a single photon. This discovery established that the photon consists of two fields of energy, each of which can be transformed into an elementary particle of matter.

A quantum is presumptive matter and antimatter

Pair particle production established that each photon consists of a presumptive particle of matter and antimatter. Because antimatter does not exist in the normal world of matter, it made sense that the presumptive antimatter field of energy in a quantum would not react with matter.

Matter-antimatter annihilation

Anderson confirmed Shuster's speculation and Dirac's prediction that when matter encounters antimatter, mass is annihilated. The product of annihilation is generally two gamma ray photons moving off in opposite directions.

Clauser - Freedman experiment

The gamma ray photons generated in matter-antimatter annihilation were used in an experiment conducted at Berkeley in 1972, by John Clauser and Stuart Freedman who arranged for the annihilation photons to pass through polarizing filters and into photomultipliers to record them. When they inverted one photon, they found that the other one flipped-over as well so that they were both plane polarized at the same time. [2]

2   **Clauser** J. F. & **Freedman** S. J. *Experimental test of local hidden-variable theories*, Phys. Rev. 1972

The Clauser - Freedman experiment appears to confirm the photon as a long entity. If the 'tips' of the gamma ray photons en-

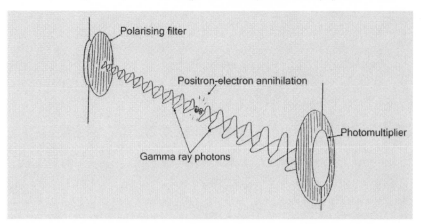

tered the polarizing filters before their 'tails' had left the site of generation, the flip of one photon would influence the second photon leaving the same point of generation in the opposite direction. This account for the Clauser - Freedman experiment supports the wave-train model for the photon of light.

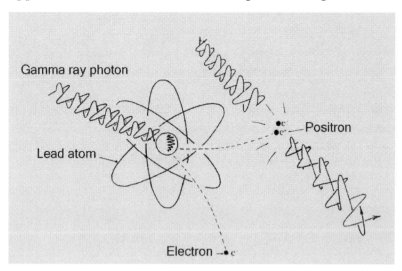

Antimatter and the Quantum laws of Motion

Carl Anderson found that the positron and electron pairs are produced only if the gamma ray photon encountered the nucleus of an atom.[3] Each of the two wave trains of light in the gamma ray photon, driving into the proton and neutron vortices in an atom of lead would have been transformed into vortices in accordance to the first quantum law of motion.

The two subatomic vortices appeared on the other side of the nucleus as the electron-positron pair. They emerged with a sum mass equivalent to the energy of the original gamma ray photon according to $E=mc^2$. As the positron lost its kinetic energy, it was drawn toward an electron by opposite charge attraction culminating in their mutual annihilation. This is explained in the vortex theory as two vortices of equal size, but opposite direction of spin effectively unzipping one another; destroying not the energy but the vortex form of energy.

In line with the second quantum law of motion the two lines of the movement of light reverted to their original waveform and radiated away as gamma ray photons. A fascinating feature of this operation of the second quantum law of motion is that the electron annihilated was not the new electron that came into existence but an old one that had been around for billions of years.

At the end of the process the energy accounts were settled as the photons produced by annihilation carried away the energy of the particle pair provided by the original gamma ray photon. This experiment provided vivid confirmation that energy conservation laws apply only in the overall process of production

3   **Richards et al,** Modern University Physics, Addison Wesley 1973

and decay of unstable particles, not in the detail of the steps in between.

Symmetry Stress

The annihilation of antimatter corrects a stress in universal symmetry set up by its production. In the cosmology of the vortex theory if there is a symmetrical world of antimatter beyond the centre of matter - which acts as a source and sink of energy flowing through vortices in matter – every form and action in matter would have to be duplicated in antimatter to satisfy the symmetry. When a particle of antimatter pops up in our world a particle of matter would appear in the world of antimatter. This reversed vortex pair would have stressed the symmetry of the worlds of matter and antimatter. By annihilating with a normal pair the symmetry stress of vortex particles flipped into the wrong worlds would be corrected.

Predictions of an antimatter half of the Universe

Speculations of an antimatter half to the Universe are supported by the law of conservation of charge formulated by Lord Rutherford and Fredrick Soddy. The law directed that for every particle of matter created an equal particle of antimatter is created and for every particle of matter annihilated an equal particle of antimatter is annihilated.

Ideas of an antimatter world were rejected for lack of evidence of gamma rays that occur with matter-antimatter annihilation. Instead the speculation that when the Universe began with a big bang there was more matter than antimatter has been generally accepted. The idea that antimatter then annihilated most of the matter leaving the remnant which we observe as the current universe is an example of a limited belief and speculation should be dismissed if they lack immediate evidence.

The limitation of 'seeing is believing'

The generally accepted outlook that 'seeing is believing' and speculations should be dismissed if they lack immediate evidence, has serious limitations. Rather than allowing the possibility that evidence of an antimatter half of the Universe might be found eventually, scientists denied the possibility. When gamma ray bursts from galactic cores were discovered, that could only be accounted for in terms of matter-antimatter annihilation, scientists have been left embarrassed having endorsed a limiting belief based on denial. .

The Quantum Loop

In the vortex theory, vortex energy is thought to circulate between subatomic particles of matter and antimatter connected at a common point of singularity at their centres.

This gives a picture is reminiscent of the infinity sign. The endless flow of vortex energy between the connected loops of matter and antimatter set up the infinity of space, and the infinite extension of force fields. Symmetry suggests that energy should flow

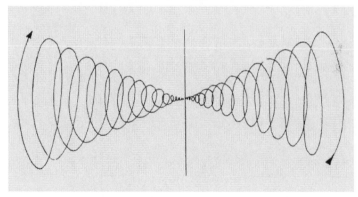

in a loop in the wave form quantum as well as the vortex pair. Energy could flow through a photon in two directions in a *quantum loop* to set up an infinite cycle in the quantum between the worlds of matter and antimatter.

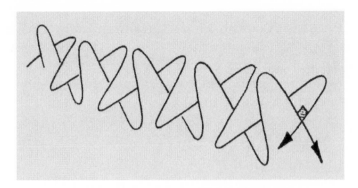

Size of space is a feature of the Vortex.

The connection between matter and antimatter, at either end of the dimension of size is a feature of the vortex. The size of space between the extremes of largeness and smallness are a consequence of the vortex. Space only appears to be immense relative to human perspective. Distances between galaxies in deep space appear to be vast to us and distances between subatomic particles in the atom appear miniscule because we measure relative to our own size. Taking into account the size of galaxies and atomic nuclei the distances in atomic space and deep space could be viewed as comparable if size is taken relative to position in the dimension of size of space.

The size of space doesn't apply to the quantum loop

The wave quantum of energy is not a vortex. It is not a three dimensional extension in the dimension of size therefore the size of space does not apply to it. The fields of presumptive matter and antimatter in the quantum loop would not be separated in the extension of space but occur in parallel on its length each in the space of matter or antimatter.

Quantum Loop Symmetry.

The separation between matter and antimatter into two distinct worlds is a feature of the vortex form of energy. As it does not apply to the radiant form of energy a quantum could exist in both worlds simultaneously interacting with the electromagnetic fields in atoms of matter or antimatter in the mirror worlds at the same time. As the quantum drives into a vortex in the world of matter it would be driving into a vortex of antimatter so that the actions are replicated in both worlds. This is called *Quantum Loop Symmetry*

The Reverse Time Paradox

Time is the flow of energy in one system relative to another. The directional flow of time from past, through present into future in our experience is the flow of energy in the quantum, in the field that interacts with matter. The reverse flow of energy in the other field would set up reverse time if it reacted with matter which fortunately it doesn't. Backward running time would set up impossible time paradoxes.

Antimatter corrects time reversal.

Time is reversed with antimatter. This is another reason why the idea of an antimatter world was unpopular. The reverse flow of energy in the quantum, setting up a time reversal, would be in the world of antimatter where time is already reversed. Two minuses make a plus. The two reversals would correct each other so that in the world of antimatter the flow of energy in the quantum would be a forward directional flow of time from past through present and into the future. That would avoid impossible time paradoxes and secure symmetry between the two worlds. Reverse time set up by matter-antimatter production is a symmetry stress that is corrected by matter - antimatter annihilation.

Quantum Loop Reverse Symmetry & Neutrinos

Quantum Loop Reverse Symmetry occurs when a quantum loop is reversed in the matter-antimatter Universe. Reverse symmetry occurs with the vortex in pair particle production so it could also occur with the quantum. The field that interacts with matter would be in the world of antimatter and the field that interacts with antimatter would be in the world of matter. In neither reality would the wave energy in the quantum interact with the vortex. Energy would be conserved in both cases but through lack of interaction the energy would be beyond detection. Quantum loop reverse symmetry could account for the non-interacting neutrino.

Appendix 7: Relativity

Understanding Einstein's relativity

Einstein's key principle in relativity is that: *The observed velocity of light is independent of the velocity of the observer.* In the vortex theory there is a very simple explanation for this. If the vortex energy extending from a body of matter is space then this 'bubble of space' would move with the body. An observer can only measure the speed of those photons of light, which reach him. As they do so they would be traveling in the bubble of space extending from him. Because this moves with him wherever he goes, his measure of the speed of light would be independent of his velocity of movement toward or away from the source of light.

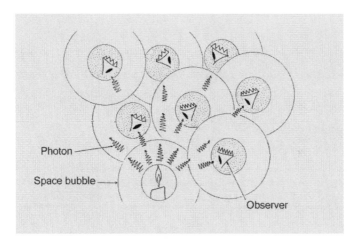

To understand this idea imagine you were in a car measuring the speed of a swarm of bees. If you were to drive alongside the bees until there is no relative movement between you and them, their speed would be the same as yours. You could put your foot on the accelerator and race away from them.

If you were to turn round and head back for the swarm, as you pass them in the opposite direction at the same speed, they would appear to be flying past very fast. You would notice the bees hit the windscreen harder when you drive back into the swarm than when you were driving along with it. However, angry bees not take kindly to experiments of this sort, even in the name of science, might crawl through gaps and find their way into your car.

Your next discovery would be that the velocity of the bees in the car would be independent of the velocity of the car because no matter how hard you drive, in any direction, you wouldn't escape their stings. In this analogy, the car represents your own space bubble extending from you and the bees represent the photons of light whose speed you are measuring. The photons in your space would be moving with you. Their velocity would therefore be independent of yours. They are represented by the bees in your car.

The Michelson and Morley experiment.

When Michelson and Morley conducted their classic experiment in 1887 measuring the velocity of light relative to the movement of the Earth, they found that the measured velocity of light was independent of the velocity of the Earth. According to the vortex theory Michelson and Morley were measuring the velocity of light relative to the space extending from the Earth. That space was moving with the Earth. That is why their measure of the velocity of light was not affected by the movement of the Earth.

Space foreshortened

Einstein believed space was foreshortened by velocity. From the consideration of relative velocity, we might expect that light emitted from a body, approaching an observer at 100,000 miles per second would appear to have a velocity of 186,000 + 100,000 =

286,000 miles / second. But Einstein's special theory tells us that the 'mile' on a body moving at 100,000 miles per second is shorter than a stationary mile so that the light does not travel as 'far' in a second from the moving body and its velocity still works out at 186,000 miles per second.

The account for relativity in the vortex theory is more congruent with common sense than Einstein's. While the vortex theory is fundamentally different from the special theory of relativity, Einstein's concept that space contracts with acceleration does work with the vortex by the following argument.

If space were created by the vortex, then the size of space would be a function of the size of the concentric spheres of energy forming the vortex. Near its centre these would be very small but far from the centre of the vortex the spheres would be large. The result of this would be that acceleration down the radius of the vortex, toward its centre, would be accompanied by a contraction of space.

A Fourth Dimension of Space

The contraction or expansion of space sets up the fourth dimension of space we experience when we grow from a single cell into adulthood. This is the dimension of size which could be called the 'Fractal Dimension' because fractals are repeating patterns in the dimension of size. In the vortex theory, I named the dimension of size the 'Alician Dimension' after Alice from *Alice's Adventures in Wonderland*.[1]

The Illusion of Forms

A feature of the Alician or Fractal Dimension is the illusion of forms which occurs with both matter and space. As energy extends out of the vortex the inverse square law sets an intensity at any given distance from the centre. Vortex energy constantly

1   **Carroll** Lewis, Alice's Adventures in Wonderland, 1865

189

flows through each point but because the intensity is always the same at that point it sets up the illusion of stasis. For example the form of your body appears to be the same but the vortex energy flowing through you is constantly changing from one moment to the next.

The illusion of forms is apparent in rivers and streams. In a river or a stream the form of a feature in the water will appear to be the same even though the water flowing through it is constantly changing. An analogy for the illusion of forms is a large family. Children represent the energy flowing through the illusory form of the vortex represented by the family. As each child grows a year older there is a lesser sibling growing behind him or her to take over the age. The family appears stable and enduring because each age position is filled by a child every year until the births eventually stop.

When the children all grow up and leave home the family breaks up. The children then repeat the pattern creating families of their own. The illusion of forms, illustrated by the family, is a feature of all forms of matter and space. It is the cause of *maya* the illusion set up by the vortex of energy. Nothing lasts. Forms are lost and though patterns may be renewed, everything is eroded by time and time is the flow of energy through the illusion of form set up by the vortex.

Space Shells

I call the levels of intensity of vortex energy that set up illusory forms *shells of space*. The shells of space set up the stable form of a body of matter. The flow of vortex energy through shells of space sets up time. As the shells of space shrink with acceleration toward the centre of the vortex, time dilates.

Einstein's General theory of Relativity

In his general theory of relativity, Einstein used the dilation of time with acceleration in an ingenious way. He suggested that if acceleration led to time dilation then the dilation of time should appear as acceleration. He used this idea in his account for gravity. He suggested that space and time formed a single, four-dimensional continuum. Then he suggested that mass distorts the space-time; the distortion manifesting as a curvature of space and a dilation of time. Einstein used the image of a sheet of rubber as space-time and heavy balls as massive bodies of matter. [2]

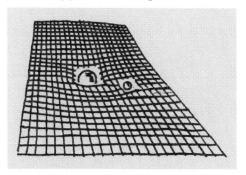

In the rubber sheet model, the ball stretches the rubber and creates a dip. The stretching of the sheet represents the stretching of time and the dip depicts the curvature of space. As another ball enters the dip it rolls in the curve toward the first ball. Einstein's view of gravity was the acceleration of the second body of matter toward the first in the curvature of space-time.[4] Einstein suggested that this gravitational effect occurred not because the object entered a gravitational force field, but because it entered a region in which space-time was distorted by a massive body of matter. The result of entering a region where time slowed-down would be that the object appears to accelerate. Einstein indicated that the energy of acceleration would come from the destruction of mass. He dismissed the 'force field' theory for gravity on the grounds that it is impossible to show by experiment whether it was a force field or acceleration that caused the gravitational attraction. He called this the *principle of equivalence of forces.*

2   **Calder** Nigel, *Einstein's Universe*, BBC Publications, 1979

The Principle of Equivalence of Forces

Einstein's principle of equivalence of forces is easy to understand. Occupants of a rocket ship experience a fall toward the rear as it accelerates away from the Earth but if the rocket were to fall back toward the Earth its occupants would experience weightlessness. Einstein argued that the occupants of the rocket ship would have no way of distinguishing between the force of gravity and the acceleration of their ship. He concluded that likewise we on Earth could not be certain whether gravity was the result of acceleration or the action of a force.

Many people believe that Einstein toppled Newton with his general theory of relativity as he appeared to have dismissed Newton's view of gravity. However, the principle of equivalence of forces does not allow this. The principle does not allow for the acceleration theory to be more correct than the force theory. It requires equivalence for the two *points of view*.

The principle requires that both Newton and Einstein are correct in their views on gravity. If the principle of equivalence is valid

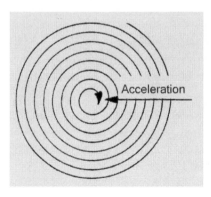

then a complete theory for gravity has to encompass the opinions of Newton and Einstein, and this is what the vortex theory does. A vortex of energy is motion in a spiral so relative to another vortex it is acceleration. This is because motion in a spiral results in acceleration down the radius of the spiral. A spiral of tape on a tape recorder illustrates this. As the tape winds off the spool, at a uniform speed, the spool accelerates.

If two spheres are placed in contact they touch at the point of a radius. Spherical vortices will always connect with each other along a radius therefore each will be relative to the other on acceleration. Because of this, radial interactions of the vortex can be treated as relative accelerations as well as interacting force fields. It all depends upon the observer's point of view.

In the vortex theory Einstein's principle of equivalence of forces applies to the force of electric charge as well as gravity which makes sense if they are both treated as vortex interactions and share the same equations. In the interactions of electric charges, if two vortices were viewed from above, they would appear as overlapping force fields.

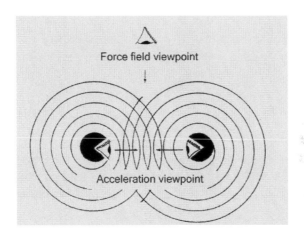

On the other hand, if the vortices were viewed relative to one another along their radii then each would appear to the other as acceleration. Either point of view would record the same results. It is impossible to design an experiment which will show one point of view to be more valid than another, therefore, it would be impossible to show by experiment whether the force of charge was the result of relative accelerations or interacting force fields. The application of the principle of the equivalence of forces to both electric charge and gravity confirms that they are in essence the same.

The Vortex Theory unifies Electromagnetism & Gravity

Quantum Electro Dynamics does not conform to Einstein's principle because it is based on Heisenberg's principle; and they are irreconcilable which is why, as already noted, Stephen Hawking commented, *"The main difficulty in finding a theory that unifies gravity with the other forces is that general relativity is a classical theory in that it does not incorporate the uncertainty principle of quantum mechanics."*

It has proved impossible to unify the forces of nature through quantum mechanics. Gravity refuses to fit. In the vortex theory, by contrast, it has proved possible to reconcile the force of gravity with electromagnetism because the vortex force of electric charge complies with Einstein's principle rather than Heisenberg's.

Resolving the Newton Dilemma

The vortex theory makes it possible to overcome a dilemma raised by Isaac Newton in his endeavor to explain action at a distance. In a letter to his friend, Richard Bentley, Newton wrote: *"...It is inconceivable that inanimate brute matter should, without the mediation of something else which is not material, operate upon and affect other matter without mutual contact..."*

The vortex model deals with the Newton dilemma by depicting particles of matter as neither inanimate nor brute. Because of the infinite extension of the vortex of energy, particles of matter are never out of mutual contact. Vortex interactions remove the dilemma raised by assuming the existence of a mediating, non-material force to account for gravity because the vortex is both the particle and the force field extending from it.

When Newton conceived of gravity he was sitting in an orchard pondering on the fall of an apple. Prior to its fall the apple wasn't moving, it was attached to a tree. Something must have been pulling on it to cause it to become detached and then move toward the Earth. Newton argued that the same pull must have been acting on all bodies of matter to draw them together. He suggested that this pull would explain the orbits of the moon and planets. Newton's theory was very successful but not as accurate as Einstein's. Newton's theory could not account for the precession of planetary orbits – most notable in the orbit of Mercury – but this was because Newton's theory did not allow for the curvature of space. Gravity acting through curved space provides a full account for planetary orbits.

The Vortex account for curved space

The vortex theory provides an account for curved space but this is different from Einstein's. Einstein based his theory for gravity on the assumption that matter distorts space-time.

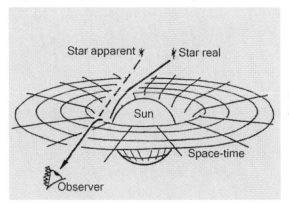

In the vortex theory space is curved not because it is distorted by matter, space is curved because it is an extension of matter. The vortex explanation for the curvature of space shows the same effects on light as Einstein's description. Einstein predicted that the distortion of space-time around the sun would cause an apparent curvature of starlight as it passes through the 'dip'. The vortex theory shows that starlight would follow a curved path as it passes through the spherical shells of space extending from the sun much as a car follows a curved path around a roundabout.

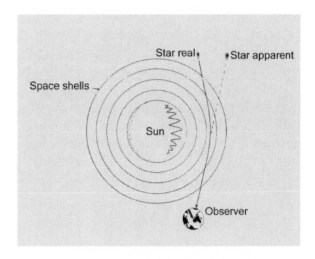

The vortex theory depicts light flowing round the sun like a car driving round a roundabout with the car going round because the road goes round. This simple account of space curvature has the advantage of not assuming time to be a fourth dimension in continuum with space. In the vortex theory space has a fourth dimension; the dimension of size of space. It makes more sense for the fourth dimension of space to be a feature of space not time.

The weakness of Newton and Einstein's theories

The weakness in both Newton and Einstein's theories for gravity of is that they both contain arbitrary assumptions. To explain gravity Einstein assumed that space and time exist as a continuum that is distorted by mass. Newton assumed that gravity was caused by something non-material acting between bodies of matter. Einstein never gave us a clear picture of what space-time was that it could be distorted by matter and Newton never left us a clue as to what the mysterious non-material entity was that acted between bodies of matter to pull them together.

The strength of the Vortex theory

The strength of the vortex theory is it accounts for gravity from a unified description of space and mass as aspects of the vortex of energy. Because the vortex is formed of movement at the speed of light, space, time and mass are relative to this invariable speed. Because the quantum vortex is curved, space is curved. Because vortex energy extends beyond our direct perception, bodies of matter can act on each other at a distance. Because the quantum vortex is intrinsically dynamic these interactions can occur without the mediation of any non-material entities. There are no arbitrary assumptions laid on the single axiom that sub-atomic particles are vortices of energy. The forces and properties of matter including gravity and space are accounted for in a single uninterrupted flow of unfolding logic.

The common properties of space and gravity

Einstein treated gravity as a property of space and gravity has a lot in common with space. However, it is important to appreciate that like matter, space is static because it represents the levels of intensity of energy through which energy flows rather than the flow of energy itself.

Gravity	Space
1: Extends from matter	1: Extension of matter
2: Acts from centre of each particle of matter	2: Extends from the centre of each particle of matter
3: Unlimited range in 3D	3: Unlimited extension in 3D
4: Obeys inverse square law	4: Obeys inverse square law

The graviton account for gravity is unsatisfactory

The idea in quantum mechanics that gravity is caused by the exchange of virtual particles called *gravitons* is unsatisfactory because it is hard to imagine that the exchange of force carrying particles within the bounds of Heisenberg's uncertainty principle could account for the instantaneous action of gravity over the expanse of deep space.

Quantum mechanics is a classical theory

Quantum mechanics is a classical theory because it is based on the assumption that particles exist with predetermined properties then move through space. If Quantum theory were hinged on Einstein's discovery that movement at the speed of light is the constant to which everything is relative including mass, space and time and the properties of particles were seen to arise from the forms of motion and their relative interactions, quantum theory could then incorporate relativity instead of the classical atomic hypothesis.

Appendix 8: Super-energy

Beyond absolutism in Einstein's relativity

Before the turn of the 20th century, mass, space, time and the laws of physics were thought to be absolute and invariable.

However, early in the 20th century, it became clear that these classical assumptions were invalid. Einstein showed that space and time, mass and the laws of physics were not absolute and invariable but were relative to the speed of light. In the 20th century, it was the turn of the speed of light to be treated as absolute and invariable. Now in the 21st century this final absolutism can be questioned. It is time to move beyond the absolutism in Einstein's relativity.

Why particles don't go faster than the speed of light

Sub-atomic vortex particles cannot be accelerated beyond the speed of light when they are accelerated by wave-particles of energy moving at the speed of light. If a moving particle is a vortex propelled by wave trains of energy it would be accelerated by the input of additional quanta of energy, overcoming its static inertia by their kinetic inertia. If the propelling packets of energy are particles of movement at the speed of light obviously this would be the maximum speed any particles propelled by them could ever achieve. This does not mean that the speed of light is absolute and invariable throughout the Universe. Nor does it prove the speed of light to be the absolute limit of speed in the Universe. Certainly it does not prove all energy is constrained by the speed of light.

Why energy could exist beyond the speed of light

If energy is pure movement it is entirely possible that there could exist elsewhere in the Universe particles of movement with intrinsic speeds faster than that of light. The same laws of physics we observe could apply to these particles of energy. They could occur in the form of vortex and wave. Only their relativity constant would be different. Their mass, space and time and physical laws pertaining to them would be relative to a constant beyond the speed of light.

Super-energy

I predict the existence of energy beyond the speed of light. I call is and predict that particles of super-energy, in the forms of wave and vortex, could set up worlds of atomic and non-atomic matter. These worlds could teem with life. They might form parallel universes and include planets and stars. The only difference between these worlds and our own would be the speed value of the constant of relativity. Vortices and waves of super-energy existing with different intrinsic speeds could set up a range of realities beyond the speed of light. This prediction would allow for the possibility of super-energy establishing a number of planes of space and time distinct from our own.

Distinct realities

The laws of relativity physics do not allow particles formed of energy with speeds faster than that of light to exist within a space-time continuum relative to the speed of light. Waves and vortices of super-energy could, however, establish their own space and time because space and time are an aspect of the vortex. Differences in the speeds of energy and super-energy would set up a distinction in the realities they form. The distinction would establish a dimension between them.

A dimension of speed

The dimensions of space time and size separate things in our world. The distinction separating our world of energy from a world of super-energy would be a dimension based on the differences in their intrinsic speeds of energy. The dimension of speed would be discontinuous, that is it would occur as distinct steps, each representing a particular speed of energy pertaining to a different level of *quantum reality*. Much as particles in the atom occur with distinct masses and quantum states in the dimension of speed energies could occur with distinct intrinsic speed values. This would lead to a series of levels or planes of reality in the Universe; each dependant on a different intrinsic speed of energy.

The cosmic law of motion

There is a self evident law of motion that dictates: *Lesser speeds are subsets of greater speeds*. This suggests the energy on each level in an ascending series would be a subset of the level above it. Because slower speeds of motion are a subset or part of faster speeds, realms of energy based on lesser speeds would be a subset or part of realms of energy based on faster speeds. Applied to the cosmos this law of motion would determine that, *the world of physical-energy would be a subset of the world of super-energy*. Because space and time are relative to speed of energy, the principle of subsets would also apply to space and time. The space and time of our world, relative to the speed of light, would be a part or subset of the space and time of worlds of super-energy, based on energy beyond the speed of light.

Concentric sphere model

An ascending series of energy realities could be represented by a set of concentric spheres. Each sphere would depict an intrinsic speed of energy; a different constant of relativity. Lesser energy speeds, depicted by smaller spheres are contained by greater energy speeds shown as larger spheres.

Harmony of the Spheres

In line with the Hermetic symmetry *as it is above so it is below; as it is below so it is above*, the concentric sphere model for levels of energy in the cosmos can also be applied to the quantum states of

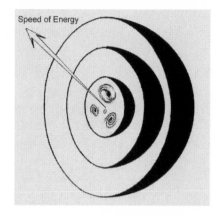

energy in the atom. The concentric sphere model for the ascending series of quantum realities appears to have been anticipated in the Hermetic concept of the *Harmony of the Spheres*. In ancient times the model of concentric spheres was used to describe the oganisation of the heavens as planes of reality The word *planet* for our world originates from this Hermetic map of the heavens; mistakenly associated with the solar system in modern times.

The position of our world in the Harmony of the Spheres, as the innermost sphere depicts a feature of speed subsets that there is no distinction within a speed i.e. all speeds below the speed of light would belong in our world based on energy at the speed of light and represented by the innermost sphere.

The distinction of speed subsets

The distinction in speed subsets is up not down. Every speed contains all lesser speeds as part of its continuum of motion but it does not contain greater speeds. A lesser speed may be part of a greater speed but the greater speed is not part of it. Every bit of energy, every speed of motion between zero and the speed of light is part of our world and relative to the speed of light, but not super-energy. Super-energy would not be part of our world. We would not see it, feel it or measure it with our scientific apparatus because it would not be within our continuum of space and time. Our light and the forces in our world based on energy at the speed of light would not interact with super-energy beyond the speed of light any more than a bloke on a bike could catch a jet.

Though super-energy would be invisible and intangible to us we would not be so to it. Were worlds of super-energy to exist our world would be a part of them. Waves and vortices of super-energy could, therefore, interact with waves and vortices of physical energy. If there were sentient beings in the realms of super-energy, though we could not see or feel them, they would be able to see and feel us. That is how subset distinction works.

Sentient beings in super-energy

We are sentient beings. If our world is part of a greater world of super-energy it is obvious there could sentient beings in super-energy. To presume otherwise would be a logical absurdity. Like a blind man lording it in his hovel ignorant of a king in the palace beyond, so the greatest in the lesser part could be the least in the greater whole. The law of subsets is everything in the lesser part is possible in the greater because the lesser is part of the greater.

The Vortex Universe

The conical vortex provides a model for the Universe.

The line in the conical vortex represents speed of energy. Each complete turn in the spiral represents a critical speed of energy acting as the relativity constant. Each plane is a quantum reality in an ascending series of realities. The first turn at the base of the spiral would represent the speed of light, the relativity constant for our physical world. The base of the vortex represents the known reality of matter and light. The progressively higher turns in the spiral represent possible planes of reality in super-energy, each relative to a higher speed of energy beyond the speed of light.

Parallel Vortex Universes

There could be a set of parallel vortex universes. Allowing for the  principle of fractals a set of parallel vortex Universes could exist as part of a greater system much like the arrangement of dandelion seeds in a flower head. Then this could be but one amongst many sets, just like the countless dandelions bobbing in the breeze in spring time.

Appendix 9:

Continuous Living

Death is NOT the end

Dr. Sam Parnia[1], Director of the AWARE Study (Awareness during Resuscitation) is one of the world's leading experts on the scientific study of death and near-death experiences. He is at the cutting-edge of research at the front line of critical care and resuscitation medicine. In his book Erasing Death: The Science That Is Rewriting the Boundaries Between Life and Death 1, Sam Parnia makes it clear we are on the verge of discovering a new universal science of consciousness that reveals the nature of mind and a future where, contrary to popular belief, death is not the end but a transition in a continuum of living.

Revolutionary resuscitation procedures

In his revolutionary resuscitation procedures involving the suspension of the body after it has died, now employed in hospitals throughout the United States and Europe, patients are returning from death with experiences life after death that match reports from near death experiences. This scientific research is establishing beyond doubt that there is continuity of human consciousness during and after death. Dr. Parnia's research is revealing evidence of continuous living; that is the continuation of the human mind and psyche after the brain stops functioning and the body dies. Novel resuscitation techniques proven to be effective in revitalizing both the body and mind after death are providing mounting evidence of the

1   Parnia S., Erasing Death Harper and Collins 2014

universality of consciousness and a spiritual existence after the physical.

How much more proof do we need?

Innovative techniques, such as drastically reducing the patient's body temperature, are now used routinely by doctors to reverse death. New medical discoveries focused on saving lives are inadvertently answering questions about the self and the soul. Questions that were once relegated to theology and philosophy are now being examined in rigorous scientific research. There is no longer a need to prove the existence of universal consciousness, life-after-death, soul, spirit and life beyond the limited world in which we live. How much more proof do we need? The need now is for a comprehensive theoretical framework of understanding in science for the overwhelming volume of evidence already in existence for continuous living.

Bibliography

Abbot A.*Ordinary Level Physics* Heinemann Education 1963

Alexander E. *Proof of Heaven* Piatkus 2012

Alexander E. *The Map of Heaven* Piatkus 2014

Amelino-Camelia G. *Double Special Relativity,* Nature 2002

Ariew R *Ockham's Razor.* University of Illinois Press 1976

Ash D *The New Physics of Consciousness,* Kima Global Publishing 2007

Ash D *The New Science of the Spirit,* The College of Psychic Studies (Queensberry Place London) 1995

Ash D & **Hewitt** P *The Vortex: Key to Future Science,* Gateway Books 1990

Ash Michael, *Health, Radiation and Healing,* Darton Longman & Todd, 1963

Berkson, William, Fields of Force: World Views from Faraday to Einstein, Rutledge & Kegan Paul 1974

Bohr Neils., Causality and Complementarity: Epistemological Lessons of Studies in Atomic Physics, Ox Bow Press 1999

Bohr N. On the Constitution of Atoms and Molecules, Philosophical Magazine, 1913

Burr H. S. *Blueprint for Immortality,* Neville Spearman, 1972

Calder Nigel, Key to the Universe: A Report on the New Physics BBC Publications 1977

Calder Nigel, *Einstein's Universe,* BBC Publications, 1979

Capra Fritjof., *The Tao of Physics,* Wildwood House, 1975

Capra Fritjof., *The Turning Point,* Fontana, 1983

Carroll Lewis, Alice's Adventures in Wonderland, 1865

Cassidy D. C., Uncertainty: The Life and Science of Werner Heisenberg, W. H. Freeman

Campbell John. Rutherford, Ernest 1871 - 1937 Dictionary of New Zealand Biography, 2007

Clauser J. F. & **Freedman** S. J. Experimental test of local hidden-variable theories, Phys. Rev. 1972

Clerk R.W. *Einstein: His Life & Times* Hodder & Stoughton 1973

Dahl F., Flash of the Cathode Rays: A History of J.J. Thomson's Electron. Institute of Physics Publishing. June, 1997

Dawkins R, *The God Delusion*, Bantam Press, 2006

Dawson J, Star Tribune of Minneapolis, May 15th 1994.

De Broglie Louis *Recherches sur la théorie des quanta* (Research on the quantum theory), Thesis, Paris, 1924.

Eddington Arthur, *Space Time and Gravitation*, Cambridge University Press 1920

Dirac **P., *The Quantum Theory of the Electron,*** Proceedings of the Royal Society of London **1928**

Einstein, Albert, *Does the Inertia of a Body Depend Upon Its Energy Content?* Annalen der Physik 1905

Einstein, Albert "On a Heuristic Viewpoint Concerning the Production and Transformation of Light", Annalen der Physik, 1905

Einstein, Albert, *Die Feldgleichungen der Gravitation* (The Field Equations of Gravitation), Koniglich Preussische Akademie der Wissenschaften 1915

Einstein, Albert *Kosmologische Betrachtungen zur allgemeinen Relativitätstheorie* (Cosmological Considerations in the General Theory of Relativity 1917 Koniglich Preussische Akademie der Wissenschaften:

Feynman Richard, The Character of Physical Law, BBC 1965

Feynman R. The Character of Physical Law, Penguin 1992

Feynman Richard (with Leighton & Sands), *The Feynman Lectures on Physics* Addison Wesley 1963

Fishman C. J. and **Meegan,** C. A. Gamma-Ray Bursts, Annual Review of Astronomy and Astrophysics 33 1995

Fritzsch H, *Quarks: The Stuff of Matter* 1983 Allen Lane.

Gamow George, Thirty Years that Shook Physics, Heinemann.

Gehrels et al *The Brightest Explosions in the Universe* Scientific American Dec 2002

Goswami A. The Self-Aware Universe, Putnam, 1995

Greenstein J & Schmidt M *The Quasi-Stellar Radio Sources.* Astrophysical J 140, 1964

Gribbin J. *In Search of the Big Bang,* Black Bantam 1986

Gribbin J. Richard Feynman: A Life in Science, Penguin 1997

Hawking Stephen, *A Brief History of Time* Bantam Press 1988

Hawking S, Black Holes and Baby Universes Bantam 1993

Heisenberg W. *Physics and Philosophy* Harper & Row 1958

Hoyle Fred, *The Intelligent Universe*, Michael Joseph, 1983

Jeans James, The Mysterious Universe, Cambridge University Press 1930

Kaku M. & Thompson J. T., ***Beyond Einstein: The Cosmic Quest for the Theory of the Universe.*** Oxford Univ Press**, 1997**

Kuhn Thomas, *Black-Body Theory and the Quantum Discontinuity: 1894-1912* Clarendon Press, Oxford, 1978

Kuhn T., The Structure of Scientific Revolutions, University of Chicago Press, 1962.

Kragh H. *Quantum Generations: A History of Physics in the Twentieth Century.* Princeton University Press 2002

Bibliography

Leggett A.J., *The Problems of Physics*, Oxford University Press 1987

Matthews R Unraveling the Mind of God Virgin 1992

Maxwell James Clerk, *Encyclopaedia Britannica*, 1875

McKenzie A. E., *A Second MKS Course in Electricity*, Cambridge University Press 1968

McTaggart Lynne, *The Field*, Harper & Collins, 2001

Nelkon & Parker, *Advanced Level Physics*, Methuen 1982

Pagels Heinz, The Quantum Code, Michael Joseph 1982

Pasachoff Naomi *Marie Curie and the Science of Radioactivity*, New York, Oxford University Press, 1996.

Planck, Max. "On the Law of Distribution of Energy in the Normal Spectrum". Annalen der Physik, vol. 4, p. 553 ff.1901

Pond D *Universal Laws* Infotainment Books, 1995

Popper Karl, *The Logic of Scientific Discove*ry Hutchinson 1968

Ramacharaka Yogi, An Advanced Course in Yogi Philosophy Fowler 1934

Rees Martin, *Just Six Numbers* Weidenfield & Nicolson 1999

Richards, Sears, Wehr & Zemansky, Modern University Physics, Addison Wesley 1973

Russell B. *The ABC of Relativity*, George Allen & Unwin, 1958

Russell P From Science to God: Exploring the Mystery of Consciousness New World Library, 2005

Samanta-Laughton M. *Punk Science* O Books 2006

Sartori P. The Wisdom of Near Death Experiences Watkins 2014

Sartori P. The Near Death Experiences of Intensive Care Patients: A Five Year Clinical Study Edwin Mellen Press 2008

Schroeder G. *The Hidden Face of God* Free Press 2001

Schrödinger E *Collected papers* Friedr.Vieweg & Sohn 1984

Schrödinger E, *Science and Humanism* Cambridge University Press 1951

Schuster *A. Potential Matter: A Holiday Dream*, Nature (1898).

Segrè Emilio, *Nuclei & Particles* Benjamin Inc 1964

Segrè Gino, Faust in Copenhagen: A Struggle for the Soul of Physics, Viking 2007

Sheldrake Rupert, *A New Science of Life*, Paladin Books, 1987

Thomson J.J., *Treatise on the Motion of Vortex Rings* University of Cambridge 1884

Thomson, S.P., Life of William Thomson, Baron Kelvin of Largs 1910

Thomson, William: *Mathematical and Physical Papers* 6 vols. 1841-1882

Thomson, William; Popular Lectures and Addresses

Tiller W. Science and Human Transformation Pavior, 1997

Velikovsky I. *Worlds in Collision* Victor Gollancz Ltd., 1950

Wolf F.A. *Starwave* Macmillan 1984

Zukav G The Dancing Wu Li Masters, Rider 1979

Index

Index

Index

Universe 8, 10 - 12, 14, 16, 25 - 28, 31, 327, 47, 50, 63, 75, 83, 97 - 99, 101, 111 - 116, 120, 123, 124, 129, 131, 182, 203, 204

V

Vector 146, 147, 164 - 168, 170 - 172

Veda 26

Vortex 16, 30, 75, 93, 122, 152, 159, 162, 168, 170, 172, 184, 194, 195, 197, 203, 204

    energy 122, 145, 1159, 164, 165, 183, 187

    model 20, 28, 30, 146, 194

    motion 10, 22, 36, 147, 172

    of energy 16, 122

    theory 194

W

Wolf, Fred 129

Wu, Chien Shiung 150

X

X-Rays 114, 157

Y

Yin and Yang 95

Yoga 12, 14, 15, 18, 104

Yogi 10, 18

    Ramacharaka 17

Yogic Philosophy 10, 12, 17, 75, 104

Young's slit experiment 156

Young, Thomas 156

Yukawa, Hideki 64

Z

Z particles 64

Zukav, Gary 125

About the Author

David Ash was born in Kent, England in 1948. His father, Dr Michael Ash, researching the link between cancer and radioactivity, introduced David to atomic physics as a boy.

David Ash started the Vortex theory in 1967 at Hastings College of Further Education and continued to develop it at Queens University of Belfast 1968/9 and Queen Elizabeth College, London University 1969-1972 and the College of St Mark and St John in Plymouth in 1973/74. On January 15th 1975 David presented his version of the vortex theory from the Faraday Rostrum at the Royal Institution of Great Britain in the auditorium where Lord Kelvin demonstrated his vortex theory for the atom a century before. David then taught physics for a number of years before his vortex theory was published (with Peter Hewitt) as The Vortex: Key to Future Science, by Gateway books in 1990 after which he set out on an international lecture tour, 1991-94. In 1995 the College of Psychic Studies published The New Science of the Spirit, which included the prediction of the accelerating expansion of the Universe subsequently confirmed by Saul Perlmutter in 1997. The vortex theory was revised and republished as The New Physics of Consciousness by Kima Global Publishing of Cape Town in 2007. The Vortex Theory is David's definitive work, updating and revising previous versions, with the addition of The Crust Slip Theory. David Ash, now living in Somerset though now single, is father of nine children and the grandfather of eleven.

Also by David Ash

Kima Global Publishers trust that you enjoyed this book and would like to introduce you to further works by our author.

The New Physics of Consciousness
Reconciling Science and Spirituality

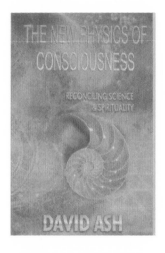

A popular and accessible presentation on physics and spirituality; a new vision of matter sits with a fresh understanding of God.

With an easy rewrite of physics there emerges a profound philosphy. Clear analogies and simple diagrams make the science understanable and enthralling. A theory for everything emerges which is simple and brilliant.

Science and religion have been reconciled. The Universe will never be quite the same again
ISBN 978-0-9802561-2-3

The Role of Evil in Human Evolution
Exposing the Dark to Light

An deeply researched investigation into the origins of evil and how it, even today, affects our entire world. The concepts in this book could be considered challenging, but are in the end extremely liberating.

ISBN 978-0-9802561-3-0

Made in the USA
Las Vegas, NV
01 December 2020